Shock and Awe:
Achieving Rapid Dominance

Shock and Awe: Achieving Rapid Dominance

Harlan K. Ullman and James P. Wade
with
L.A. "Bud" Edney
Fred M. Franks
Charles A. Horner
Jonathan T. Howe
Keith Brendley, Executive Secretary

NATIONAL DEFENSE UNIVERSITY
- President: Lieutenant General Ervin J. Rokke, USAF
- Vice President: Ambassador William G. Walker

INSTITUTE FOR NATIONAL STRATEGIC STUDIES
- Director: Dr. Hans A. Binnendijk

ADVANCED CONCEPTS, TECHNOLOGIES, AND INFORMATION STRATEGIES (ACTIS)
- Director: Dr. David S. Alberts
- Fort Lesley J. McNair, Washington, DC 20319-6000
- Phone: (202) 685-2209 • Facsimile: (202) 685-3664

Opinions, conclusions, and recommendations expressed or implied within are solely those of the authors. They do not necessarily represent the views of the National Defense University, the Department of Defense, or any other U.S. Government agency. Cleared for public release; distribution unlimited. Portions of this publication may be quoted or reprinted without further permission, with credit to the Institute for National Strategic Studies, Washington, DC. Courtesy copies of reviews would be appreciated.

Library of Congress Cataloging-in-Publication Data

Ullman, Harlan.
    Shock and awe : achieving rapid dominance / Harlan K. Ullman and James P. Wade ; with L. A. Edney ... [et al.].
       p. cm.
    ISBN 1-57906-030-7
    1. Rapid dominance (Military science) 2. Awe.
3. Shock. I. Wade, James P., 1930- . II. Edney, L. A.
III. Title.
U163.U45   1996
355.02--dc21   96-29748
                                                CIP

First printing, November 1996

For sale by the U.S. Government Printing Office
Superintendent of Documents, Mail Stop: SSOP
Washington, DC 20402-9328 Phone: (202) 512-1800

Contents

Foreword ix

Prologue... xi

Introduction to Rapid Dominance 1

Background and Basis 21

Shock and Awe..................................... 45

Strategic, Policy, and Operational Application 69

An Outline for System Innovation
and Technological Integration 105

Future Directions 137

Appendices—Reflections of Three Former Commanders

Thoughts on Rapid Dominance
by Admiral Bud Edney 145

Defense Alternatives: Forces Required
by General Chuck Horner..................... 169

Enduring Realities and Rapid Dominance
by General Fred Franks 191

Study Group Members 197

Foreword

We are in the early stages of what promises to be an extended debate about the future of conflict and the future of our defense establishment. Few will deny that the winds of change are blowing as never before, driven by a radically altered geopolitical situation, an evolving information-oriented society, advancing technology, and budgetary constraints. How our nation responds to the challenge of change will determine our ability to shape the future and defend ourselves against 21st century threats. The major issue, however it may be manifested, involves the degree of change that is required. Advocates, all along the spectrum from a military technical revolution to a revolution in military affairs to a revolution in security affairs, are making their cases. Military institutions are by their very nature somewhat conservative. History has shown that success has often sown the seeds of future failure. We as a nation can ill afford to follow in the footsteps of those who have rested on their laurels and failed to stretch their imaginations.

Often, those who are the most knowledgeable and experienced about a subject are not in the most advantageous position to understand a new world order. Yet these same individuals are often among the most credible voices and therefore are essential to progress. The authors of *Shock and Awe* are a highly accomplished and distinguished group with the credibility that comes from years of front line experience. Thus, this work is important not only because of the ideas contained within, but because of the caliber and credibility of the authors.

ACTIS seeks to articulate and explore advanced concepts. In sponsoring this work and in disseminating its initial results, we hope to contribute to the ongoing dialogue about alternatives, their promises, and their risks. As the authors note, this is a work in progress meant not to provide definitive solutions but a proposed perspective for considering future security needs and strategies. To the extent that vigorous debate ensues we will be successful.

> David S. Alberts
> Washington, D.C.
> October 1996

Prologue

The purpose of this paper is to explore alternative concepts for structuring mission capability packages (MCPs) around which future U. S. military forces might be configured. From the very outset of this study group's deliberations, we agreed that the most useful contribution we could make would be to attempt to reach beyond what we saw as the current and commendable efforts, largely but not entirely within the Department of Defense, to define concepts for strategy, doctrine, operations, and force structure to deal with a highly uncertain future. In approaching this endeavor, we fully recognized the inherent and actual limits and difficulties in attempting to reach beyond what may prove to be the full extent of our grasp.

It is, of course, clear that U.S. military forces are currently the most capable in the world and are likely to remain so for a long time to come. Why then, many will ask, should we examine and even propose major excursions and changes if the country occupies this position of military superiority? For reasons noted in this study, we believe that excursions are important if only to confirm the validity of current defense approaches.

There are several overrarching realities that have led us to this conclusion. First, while everyone recognizes that the Cold War has ended, there is not a consensus about what this means for more precisely defining the nature of our future security needs. Despite this absence of both clairvoyance and a galvanizing external danger, the United States is actively examining new strategic options and choices. The variety of conceptual efforts underway in the Pentagon to deal with this uncertainty exemplifies this reality.

At the same time, the current dominance and superiority of American military power, unencumbered by the danger of an external peer competitor, have created a period of strategic advantage during which we have the luxury of time, perhaps measured in many years, to re-examine with a margin of safety our defense posture. On the other hand, potential adversaries cannot be expected to ignore this predominant military capability of the United States and fail to try to exploit, bypass, or counter it. In other words, faced with American military superiority in ships, tanks, aircraft, weapons and, most importantly, in competent fighting personnel, potential adversaries may try to change the terms of future conflict and make as irrelevant as possible these current U.S. advantages. We proceed at our own risk if we dismiss this possibility.

Second, it is relatively clear that current U.S. military capability will shrink. Despite the pledges of the two major American political parties to maintain or expand the current level of defense capability, both the force structure and defense infrastructure are too large to be maintained at even the present levels and within the defense budgets that are likely to be approved. Unless a new menace materializes, defense is headed for "less of the same." Such reductions may have no strategic consequences. However, that is an outcome that we believe should not be left to chance.

This shrinkage also means that the Pentagon's good faith strategic reviews aimed at dealing with our future security needs may be caught up in the defense budget debate over downsizing and could too easily drift into becoming advocacy or marketing documents. As the services are forced into more jealously guarding a declining force structure, the tendency to "stove-pipe" and compartmentalize technology and special programs is likely to increase, thereby complicating the problem of making full use of our extraordinary technological resources. This means that some external thinking, removed from the bureaucratic pressures and demands, may be essential to stimulating and sustaining innovation.

Third, the American commercial-industrial base is undergoing profound change propelled largely by the entrepreneurial nature of the free enterprise system and the American personality. Whether in information or materials-related technology or for that matter in other areas too numerous to count, the nature of competition is driving both product breadth and improvement at rates perhaps unthinkable a decade ago. One sign of these trends is the reality that virtually all new jobs in this country are being created by small business. In the areas of commercial information and related management information systems, these changes are extraordinary and were probably unpredictable even a few years ago.

On the so-called information highway, performance is increasing dramatically and quickly while price, cost, and the time to bring to market new generation technology are diminishing. These positive trends are not matched yet in the defense-industrial base. One consequence of this broad commercial transformation is that any future set of defense choices may be inexorably linked to and dependent on this profound, ongoing change in the commercial sector and in learning to harness private sector advances in technology-related products. It must also be understood that only the United States among all states and nations has the vastness and breadth of resources and commercial capability to undertake the full exploitation of this revolutionary potential.

Finally, it is clear that U.S. forces are engaged and deployed worldwide, often at operating tempos as high as or higher than during the Cold War. These demands will continue and the diversity of assigned tasks is unlikely to contract. These forces must be properly manned, equipped, and trained and must carry out their missions to standards that are both high and expected by the nation's leaders and its public. The matter of maintaining this capability while attempting to reshape the force for a changing future is a major and daunting challenge not to be underestimated.

These structural realities are exciting and offer a major opportunity for real revolution and change if we are able and daring enough to exploit them. This, in turn, has led us to develop the concept of Rapid Dominance and its attendant focus on Shock and Awe. Rapid Dominance seeks to integrate these multifaceted realities and facts and apply them to the common defense at a time when uncertainty about the future is perhaps one of the few givens. We believe the principles and ideas underlying this concept are sufficiently compelling and different enough from current American defense doctrine encapsulated by "overwhelming or decisive force," "dominant battlefield awareness," and "dominant maneuver" to warrant closer examination.

Since before Sun Tzu and the earliest chroniclers of war recorded their observations, strategists and generals have been tantalized and confounded by the elusive goal of destroying the adversary's will to resist before, during, and after battle. Today, we believe that an unusual opportunity exists to determine whether or not this long-sought strategic goal of affecting the will, understanding, and perception of an adversary can be brought closer to fruition. Even if this task cannot be accomplished, we believe that, at the very minimum, such an effort will enhance and improve the ability of our military forces to carry out their missions more successfully through identifying and reinforcing particular points of leverage in the conflict and by identifying and creating additional options and choices for employing our forces more effectively.

Perhaps for the first time in years, the confluence of strategy, technology, and the genuine quest for innovation has the potential for revolutionary change. We envisage Rapid Dominance as the possible military expression, vanguard, and extension of this potential for revolutionary change. The strategic centers of gravity on which Rapid Dominance concentrates, modified by the uniquely American ability to integrate all this, are these junctures of strategy, technology, and innovation which are focused on the goal of affecting and shaping the will of the adversary. The goal of Rapid Dominance will be to destroy or so confound the will to resist that an adversary will have no alternative except to accept our strategic aims and military objectives. To achieve this outcome, Rapid Dominance must control the operational environment and through that dominance, control what the adversary perceives, understands, and knows, as well as control or regulate what is not perceived, understood, or known.

In Rapid Dominance, it is an absolutely necessary and vital condition to be able to defeat, disarm, or neutralize an adversary's military power. We still must maintain the capacity for the physical and forceful occupation of territory should there prove to be no alternative to deploying sufficient numbers of personnel and equipment on the ground to accomplish that objective. Should this goal of applying our resources to controlling, affecting, and breaking the will of an adversary to resist remain elusive, we believe that Rapid Dominance can still provide a variety of options and choices for dealing with the operational demands of war and conflict.

To affect the will of the adversary, Rapid Dominance will apply a variety of approaches and techniques to achieve the necessary level of Shock and Awe at the appropriate strategic and military leverage points. This means that psychological and intangible, as well as physical and concrete effects beyond the destruction of enemy forces and supporting military infrastructure, will have to be achieved. It is in this broader and deeper strategic application that Rapid Dominance perhaps most fundamentally differentiates itself from current doctrine and offers revolutionary application.

Flowing from the primary concentration on affecting the adversary's will to resist through imposing a regime of Shock and Awe to achieve strategic aims and military objectives, four characteristics emerge that will define the Rapid Dominance military force. These are noted and discussed in later chapters. The four characteristics are near total or absolute knowledge and understanding of self, adversary, and environment; rapidity and timeliness in application; operational brilliance in execution; and (near) total control and signature management of the entire operational environment.

Whereas decisive force is inherently capabilities driven—that is, it focuses on defeating the military capability of an adversary and therefore tends to be scenario sensitive—Rapid Dominance would seek to be more universal in application through the overriding objective of affecting the adversary's will beyond the boundaries traditionally defined by military capability alone. In other words, where decisive force is likely to be most relevant is against conventional military capabilities that can be overwhelmed by American (and allied) military superiority. In conflict or crisis conditions that depart from this idealized scenario, the superior nature of our forces is assumed to be sufficiently broad to prevail. Rapid Dominance would not make this distinction in either theory or in practice.

We note for the record that should a Rapid Dominance force actually be fielded with the requisite operational capabilities, this force would be neither a silver bullet nor a panacea and certainly not an antidote or preventative for a major policy blunder, miscalculation, or mistake. It should also be fully appreciated that situations will exist in which Rapid Dominance (or any other doctrine) may not work or apply because of political, strategic, or other limiting factors.

We realize some will criticize our focus on affecting an adversary's will, perception, and understanding through Shock and Awe on the grounds that this idea is not new and that such an outcome may not be physically achievable or politically desirable. On the first point, we believe the use of basic principles of strategy can stand us in good stead even and perhaps especially in the modern era when adversaries may not elect to fight the United States along traditional or expected lines. On whether this ability can and should be achieved, we believe that question should be part of a broader examination.

Finally, we argue that what is also new in this approach is the way in which we attempt to integrate far more broadly strategy, technology, and innovation to achieve Shock and Awe. It is this interaction and focus which we think will provide the most interesting results.

For these and other reasons, we have embarked on an ambitious intellectual excursion in making a preliminary definition of Rapid Dominance. For the moment, we view Rapid Dominance in the formation stage and not as a final product. Over the next months, we believe further steps should be taken to refine Rapid Dominance and to develop "paper" systems and force designs that will add crucial specificity to this concept. Then, this Rapid Dominance force can be assessed against five sets of questions:

- First, assuming that a Rapid Dominance force can be fielded with the appropriate capabilities of Shock and Awe to affect and shape the adversary's will, how would this force compare with and improve on our ability to fight, win, and deal with a major regional contingency (MRC)?

- Second, what utility, if any, does Rapid Dominance and its application of Shock and Awe imply for Operations Other Than War (OOTW)? Where might Rapid Dominance apply in OOTW, where would it not, and where might it offer mixed benefits?

- Third, what are the political implications of Rapid Dominance in both broad and specific applications and could this lead to a form of political deterrence to underwrite future U.S. policy? Would this political deterrence prove acceptable to allies and to our own public?

- Fourth, what might Rapid Dominance mean for alliances, coalitions, and the conduct of allied and combined operations?
- Finally, what are the consequences of Rapid Dominance on defense resource investment priorities and future budgets?

From this examination and experimentation, we believe useful results will flow.

We also would like to acknowledge the support and role of the National Defense University in sponsoring this first effort. In particular, we owe a huge debt of gratitude to Dr. David Alberts of NDU whose intelligence, enthusiasm, and wisdom, as well as his full support, have been invaluable and without which this project would have been far less productive.

Washington, D.C.

1 September 1996

L.A. Edney
C. A. Horner
H.K. Ullman

F.M. Franks
J.T. Howe
J.P. Wade

Introduction to Rapid Dominance

The military posture and capability of the United States of America are, today, dominant. Simply put, there is no external adversary in the world that can successfully challenge the extraordinary power of the American military in either regional conflict or in "conventional" war as we know it once the United States makes the commitment to take whatever action may be needed. To be sure, the first phase of a crisis may be the most difficult—if an aggressor has attacked and U.S. forces are not in place. However, it will still be years, if not decades, before potential adversaries will be able to deploy systems with a full panoply of capabilities that are equivalent to or better than the aggregate strength of the ships, aircraft, armored vehicles, and weapons systems in our inventory. Even if an adversary could deploy similar systems, then matching and overcoming the superb training and preparation of American service personnel would still be a daunting task.

Given this reality that our military dominance can and will extend for some considerable time to come, provided we are prepared to use it, why then is a re-examination of American defense posture and doctrine important? The answers to this question involve (1) the changing nature of the domestic and international environments; (2) the complex nature of resolving inter and intra-state conflict that falls outside conventional war, including peacekeeping, and countering terrorism, crime, and the use of weapons of mass destruction; (3) resource constraints; (4) defense infrastructure and technical industrial bases raised on a large, continuous infusion of funding now facing a future of austerity; and (5) the vast uncertainties of the so-called social, economic, and information revolutions that could check or counter many of the nation's assumptions as well as public support currently underwriting defense.

It is clear that these so-called grey areas involving non-traditional Operations Other Than War (OOTW) and law enforcement tasks are growing and pose difficult problems and challenges to American military forces, especially when and where the use of force may be inappropriate or simply may not work. The expansion of the role of UN forces to nation-building in Somalia and its subsequent failure comes to mind as an example of this danger. It is also arguable that the formidable nature and huge technological lead of American military capability could induce an adversary to move to a strategy that attempted to circumvent all this fighting power through other clever or agile means. The Vietnam War is a grim reminder of the political nature of conflict and how our power was once outflanked. Training, morale, and readiness to fight are perishable commodities requiring both a generous expenditure of resources and careful nurturing.

Thus, the greatest constraints today to retaining the most dominant military force in the world, paradoxically, may be in overcoming the inertia of this success. We may be our own worst enemy.

During the Cold War when the danger was clear, the defense debate was often fought over how to balance the so-called "strategy-force structure-budget" formula. Today, that formula has expanded to include "threat, strategy, force structure, budget, and infrastructure." Without a "clear and present danger" such as the Axis Powers in 1941 or, later, the Soviet Union to coalesce public agreement on the threat, it is difficult to construct a supporting strategy that can be effective either in setting priorities or objectives. Hence, today's "two war" or two nearly simultaneous Major Regional Contingency (MRC) strategy has been criticized as strategically and financially excessive. As noted by administration officials, the current force structure does not meet the demands of the "two war," MRC strategy and, in any event, the budget will not support the planned force structure. Finally, it is widely recognized that the United States possesses far more infrastructure such as bases and facilities than it needs to support the current force, thereby draining scarce resources away from fighting power. As a result, there is a substantial defense imbalance that will erode fighting power.

In designing its defense posture, the United States has adopted the doctrine of employing "decisive or overwhelming force." This doctrine reinforces American advantages in strategic mobility, prepositioning, technology, training, and in fielding integrated military systems to provide and retain superiority, and responds to the minimum casualty and collateral damage criteria set first in the Reagan Administration. The Revolution in Military Affairs or RMA is cited as the phenomenon or process by which the United States continues to exploit technology to maintain this decisive force advantage, particularly in terms of achieving "dominant battlefield awareness." Through this awareness, the United States should be able to obtain perfect or near perfect information on virtually all technical aspects of the battlefield and therefore be able to defeat or destroy an adversary more effectively, with fewer losses to ourselves and with a range of capabilities from long-range precision strike to more effective close-in weapons.

Before proceeding further, an example is useful to focus some of the as yet unknowable consequences of these broader realities, changes, and trends. The deployment of American forces to Bosnia is a reaction to and representation of major shifts occurring in the post-Cold War world. With these shifts, this deployment is suggestive of what may lie ahead for the use, relevance, and design of military force. The legacy of Hiroshima and Nagasaki, and then, the start of the Cold War, caused the West to adopt policies for containing and deterring the broad threat posed by the Soviet Union and its ideology. Thermonuclear weapons, complemented over time by strong conventional forces, threatened societal damage to Russia. Conventional forces backed by tactical nuclear weapons were later required, in part, to halt a massive Soviet ground attack in Europe and, in part, to provide an alternative to (immediate) use of nuclear weapons.

Today, the First Armored Division, the principal American unit serving in Bosnia is, in essence, the same force that fought so well in *Desert Storm* and, for the bulk of the Cold War along with our other units, had been designed to defend NATO against and then defeat a numerically superior, armored and mechanized Soviet adversary advancing across the plains of Germany. Now these troops, as well as others from both sides of the former Iron Curtain, are engaged in OOTW for which special training, rules of engagement, command arrangements, and other support structures have been put in place at short notice, few of which were even envisaged a few years ago. These are also operations that, because of intense, instantaneous media coverage, can have huge domestic political impact especially if events go wrong.

Whether or not this armored division is the most optimally configured force for such an operation is not relevant for the moment even though this unit probably was the most appropriate for this task. However, it is prudent to examine the consequences of changing tasks presaged by Bosnia, in which the enemy is instability rather than an ideological or regional adversary we are trying to contain or defeat and neutrality on our part may be vital to the success of the mission. Do these changes mean that we should alter our traditional approach to the doctrine for and design of forces? If so, how? Are there alternative or more effective ways and means to conduct these peacekeeping-related operations? And, in this evaluation, are there alternative doctrines we should consider to fight wars more effectively as we envisage scenarios under the construct of the MRC?

With the end of the USSR and absent a hostile Russian superpower, there is no external threat to the existence or survival of the United States as a nation and there will not be such an immediate threat for some time to come. This means that there is a finite window of opportunity when there is no external adversary threatening the total existence of American society; that our forces are far superior to any possible military adversary choosing to confront us directly; and that, with innovative thought, we may be able to create a more relevant, effective, and efficient means to ensure for the common defense at the likely levels of future spending.

At the same time that the Bosnia operation is underway, the fundamental changes occurring at home and abroad must be addressed. The industrial and technical base of the United States is changing profoundly. The entrepreneurial and technical advantages of the American economy were never greater and it is small business that is creating virtually all new jobs and employment opportunities. Commercial technology and products are turning over on ever shortening cycles. Performance, especially in high-technology products, is improving and costs are being driven downwards.

Sadly, the opposite trends are still found in the defense sector, where cost is high and will create even tougher choices among competing programs, especially as the budget shrinks. Cycle time to field new generation capabilities is lengthening and performance, especially in computer and information systems, is often obsolete on delivery. The defense industrial base will continue to compress and it is not clear that the necessary level of efficiencies or increases in effectiveness in using this base can be identified and implemented, suggesting further pressures on a defense budget that is only likely to be cut.

Indeed, the question must be carefully examined of whether the military platforms that served us so well in both cold and hot wars such as tanks, fixed wing aircraft, and large surface ships and submarines represent the most effective mix of numbers, technology, strategic mobility, and fighting capability. Our national preference for "attrition" and "force on forces" warfare continues to shape the way we design and rationalize our military capability. Therefore, it is no surprise that in dealing with the MRC, American doctrine, in some ways, remains an extension of Cold War force planning. While the magnitude and number of dangerous threats to the nation have been remarkably reduced by the demise of the USSR, we continue to use technology to fill traditional missions better rather than to identify or produce new and more effective solutions for achieving military and strategic/political objectives.

While there is much talk about "military revolutions" and winning the "information war," what is generally meant in this lexicon and discussion is translated into defense programs that relate to accessing and "fusing" information across command, control, intelligence, surveillance, target identification, and precision strike technologies. What is most exciting among these revolutions is the potential to achieve "dominant battlefield awareness," that is, achieving the capability to have near-perfect knowledge and information of the battlefield while depriving the adversary of that capacity and producing "systems of systems" for this purpose.

The near and mid-term aims of these "revolutions" largely remain directed at exploiting our advantages in firepower and on fielding more effective ways of defeating an adversary's weapons systems and infrastructure for using those systems. The doctrine of "decisive or overwhelming force" is the conceptual and operational underpinning for winning the next war based largely on this force-on-force and attrition model, and winning the information war is vital to this end. Few have asked whether the pattern of employing more modern technology for traditional firepower solutions is the best one and if there are alternative ways to achieve military objectives more effectively and efficiently. In other words, can the idea of dominant battlefield awareness be expanded doctrinally, operationally, and in terms of fixing on alternative military, political, or strategic objectives?

Rapid Dominance, if realized as defined in this paper, would advance the military revolution to new levels and possibly new dimensions. Rapid Dominance extends across the entire "threat, strategy, force structure, budget, infrastructure" formula with broad implications for how we provide for the future common defense. Organization and management of defense and defense resources should not be excluded from this examination although, in this paper, they are not discussed in detail.

The aim of Rapid Dominance is to affect the will, perception, and understanding of the adversary to fit or respond to our strategic policy ends through imposing a regime of Shock and Awe. Clearly, the traditional military aim of destroying, defeating, or neutralizing the adversary's military capability is a fundamental and necessary component of Rapid Dominance. Our intent, however, is to field a range of capabilities to induce sufficient Shock and Awe to render the adversary impotent. This means that physical and psychological effects must be obtained.

Rapid Dominance would therefore provide the ability to control, on an immediate basis, the entire region of operational interest and the environment, broadly defined, in and around that area of interest. Beyond achieving decisive force and dominant battlefield awareness, we envisage Rapid Dominance producing a capability that can more effectively and efficiently achieve the stated political or military objectives underwriting the use of force by rendering the adversary completely impotent.

In Rapid Dominance, "rapid" means the ability to move quickly before an adversary can react. This notion of rapidity applies throughout the spectrum of combat from pre-conflict deployment to all stages of battle and conflict resolution.

"Dominance" means the ability to affect and dominate an adversary's will both physically and psychologically. Physical dominance includes the ability to destroy, disarm, disrupt, neutralize, and to render impotent. Psychological dominance means the ability to destroy, defeat, and neuter the will of an adversary to resist; or convince the adversary to accept our terms and aims short of using force. The target is the adversary's will, perception, and understanding. The principal mechanism for achieving this dominance is through imposing sufficient conditions of "Shock and Awe" on the adversary to convince or compel it to accept our strategic aims and military objectives. Clearly, deception, confusion, misinformation, and disinformation, perhaps in massive amounts, must be employed.

The key objective of Rapid Dominance is to impose this overwhelming level of Shock and Awe against an adversary on an immediate or sufficiently timely basis to paralyze its will to carry on. In crude terms, Rapid Dominance would seize control of the environment and paralyze or so overload an adversary's perceptions and understanding of events that the enemy would be incapable of resistance at tactical and strategic levels. An adversary would be rendered totally impotent and vulnerable to our actions. To the degree that non-lethal weaponry is useful, it would be incorporated in the ability to Shock and Awe and achieve Rapid Dominance.

Theoretically, the magnitude of Shock and Awe Rapid Dominance seeks to impose (in extreme cases) is the non-nuclear equivalent of the impact that the atomic weapons dropped on Hiroshima and Nagasaki had on the Japanese. The Japanese were prepared for suicidal resistance until both nuclear bombs were used. The impact of those weapons was sufficient to transform both the mindset of the average Japanese citizen and the outlook of the leadership through this condition of Shock and Awe. The Japanese simply could not comprehend the destructive power carried by a single airplane. This incomprehension produced a state of awe.

We believe that, in a parallel manner, revolutionary potential in combining new doctrine and existing technology can produce systems capable of yielding this level of Shock and Awe. In most or many cases, this Shock and Awe may not necessitate imposing the full destruction of either nuclear weapons or advanced conventional technologies but must be underwritten by the ability to do so.

Achieving Rapid Dominance by virtue of applying Shock and Awe at the appropriate level or levels is the next step in the evolution of a doctrine for replacing or complementing overwhelming force. By way of comparison, we have summarized how we view the differences between the doctrines of Rapid Dominance and Decisive Force in terms of basic elements that apply to the objectives, uses of force, force size, scope, speed, casualties, and technique. We recognize that there will be debate over the relative utility and applicability of these doctrines and readers are encouraged to participate.

In considering the differences between the concepts of Rapid Dominance and Decisive Force, it is important to define the terms as precisely as possible.

The goals of achieving Rapid Dominance using Shock and Awe must be compared with overwhelming force. "Rapid" implies the ability to "own" the dimension of time—moving more quickly than an opponent, operating within his decision cycle, and resolving conflict favorably in a short period of time. "Dominance" means the ability to control a situation totally.

Rapid Dominance must be all-encompassing. It will require the means to anticipate and to counter all opposing moves. It will involve the capability to deny an opponent things of critical value, and to convey the unmistakable message that unconditional compliance is the only available recourse. It will imply more than the direct application of force. It will mean the ability to control the environment and to master all levels of an opponent's activities to affect will, perception, and understanding. This could include means of communication, transportation, food production, water supply, and other aspects of infrastructure as well as the denial of military responses. Deception, misinformation, and disinformation are key components in this assault on the will and understanding of the opponent.

Total mastery achieved at extraordinary speed and across tactical, strategic, and political levels will destroy the will to resist. With Rapid Dominance, the goal is to use our power with such compellance that even the strongest of wills will be awed. Rapid Dominance will strive to achieve a dominance that is so complete and victory is so swift, that an adversary's losses in both manpower and material could be relatively light, and yet the message is so unmistakable that resistance would be seen as futile.

"Decisive Force," on the other hand, implies delivering massive enough force to prevail. Decisive means using force with plenty of margin for error. Force implies a traditional "force-on-force" and attrition approach. This concept does not exclude psychological and other complementary damage imposition techniques to enhance the application of force; they have been used throughout the history of warfare. But such non-destructive means would have an ancillary role. Military force would be applied in a purer form and targeted primarily against the military capabilities of an opponent. Time is not always an essential component. As in *Desert Shield/Storm*, enough time would have to be allowed to assemble an overwhelming force. Such a luxury is not always feasible.

The differences become clearer if broken down into their essential elements:

Elements	Rapid Dominance	Decisive Force
Objective	Control the adversary's will, perceptions, and understanding	Prevail militarily and decisively against a set of opposing capabilities defined by the MRC
Use of Force	Control the adversary's will, perceptions, and understanding and literally make an adversary impotent to act or react	Unquestioned ability to prevail militarily over an opponent's forces and based against the adversary's capabilities
Force Size	Could be smaller than opposition, but with decisive edge in technology, training, and technique	Large, highly trained, and well-equipped. Materially overwhelming
Scope	All encompassing	Force against force (and supporting capability)
Speed	Essential	Desirable
Casualties	Could be relatively few in number on both sides	Potentially higher on both sides
Technique	Paralyze, shock, unnerve, deny, destroy	Systematic destruction of military capability. Attrition applicable in some situations

Four general categories of core characteristics and capabilities have been identified that Rapid Dominance-configured mission capability packages must embrace. These are identified briefly and discussed in later chapters.

First, Rapid Dominance seeks to maximize **knowledge** of the environment, of the adversary, and of our own forces on political, strategic, economic, and military/operational levels. On one hand, we want to get into the minds of the adversary far more deeply than we have in the past. Beyond operational intelligence required for battlefield awareness, Rapid Dominance means cultural understanding of the adversary in ways that will affect both ours and their planning and the outcome of the operation at all appropriate tactical and strategic levels.

Second, Rapid Dominance must achieve **rapidity** in the sense of timeliness. Rapid Dominance must have capabilities that can be applied swiftly and relatively faster than an adversary's.

Third, Rapid Dominance seeks to achieve total **control of the environment** from complete "signature management" of both our and the adversary's information and intelligence to more discrete means to deceive, disguise, and misinform.

Fourth, Rapid Dominance aims to achieve new levels of operational competence that can virtually institutionalize **"brilliance."** In some cases, this may mean changing the longstanding principle of military centralization and empowering individual soldiers, sailors, and airmen to be crucial components in applying and directing the application of force.

As we move to turn this concept into specific doctrine and capabilities for future evaluation, there is another emerging reality to consider. If the commercial-economic sector is transforming at the current rate and breadth, it could be that, over the course of many years, the defense industrial base would follow suit, or face irrelevance and extinction. Clearly, there are certain areas in defense which will never or may never be eliminated or replaced. Nuclear systems are a current example.

Should this trend of commercial dominance play out, it may mean that military force design and procurement will become dependent on the private sector and commercial technology. Rapid Dominance is a first conceptual step to deal with this possibility.

The purpose of this paper is to outline the beginnings of the concept of Rapid Dominance, its concentration on strategy, technology and innovation, and its focus on Shock and Awe. Based on this, subsequent steps will involve expanding mission capability packages concepts consisting of operations harmonized with doctrine, organization, and systems and then move on to field prototype systems for further test and evaluation as advanced concept technology demonstrations.

Background and Basis

In both relative and absolute terms, since the end of World War II, the military strength and capability of the United States have never been greater. Yet this condition of virtual military superiority has created a paradox. Absent a massive threat or massive security challenge, it is not clear that this military advantage can (always) be translated into concrete political terms that advance American interests. Nor is it clear that the current structure and foundations for this extraordinary force can be sustained for the long term without either spending more money or imposing major changes to this structure that may exceed the capacity of our system to accommodate. As a consequence, the success of the current design and configuration of our forces may ironically become self-limiting and constraining. That is not to claim automatically that there are better military solutions or that the current defense program is not the best our political system can produce. It is to say, however, that we are well-advised to pursue alternate ideas and concepts to balance and measure against the current and planned program.

To stimulate and intrigue the reader, we note at the outset that one thrust of Rapid Dominance is to expand on the doctrine of overwhelming or decisive force in both depth and breadth. To push the conceptual envelope, we ask two sets of broad questions: Can a Rapid Dominance force lead, for example, to a force structure that can win an MRC such as *Desert Shield* and *Desert Storm* far more quickly and cheaply with far fewer personnel than our planned force both in terms of stopping any invasion in its tracks and then ejecting the invader? Can Rapid Dominance produce a force structure with more effective capacity to deal with grey areas such as OOTW?

Second, if achievable, can Rapid Dominance lead to a form of political deterrence in which the capacity to make impotent or "shut down" an adversary can actually control behavior? What are the possible political implications of this capability and what would this power mean for conducting coalition war and for how our allies react and respond?

Because Rapid Dominance is aimed at influencing the will, perception, and understanding of an

Background and Basis 23

adversary rather than simply destroying military capability, this focus must cause us to consider the broadest spectrum of behavior, ours and theirs, and across all aspects of war including intelligence, training, education, doctrine, industrial capacity, and how we organize and manage defense.

We observe at first that even with the successful ending of the Cold War, the response of the United States in re-evaluating its national security and defense has been relatively and understandably modest and cautious. In essence, while the size of the force has been reduced from Cold War levels of 2.2 million active duty troops to about 1.5 million, and the services have been vocal in revising doctrine and strategy to reflect the end of the Soviet threat, with the exception of emphasis on jointness, there are few really fundamental differences in the design and structure of the forces from even 10 or 15 years ago.

Throughout the Cold War, the defense of the United States rested on several central and widely accepted and publicly supported propositions. The "clear and apparent danger" of the Soviet threat was real and seen as such. The USSR was to be contained and deterred from hostile action by a combination of political, strategic, and military actions ranging from the forging of a ring of alliances surrounding the USSR and its allies to the deployment of tens of thousands of nuclear and thermonuclear weapons.

Following the truce ending the Korean War, a large, standing military force was maintained and defined by the operational requirements of fighting the large formations of military forces of the USSR and its allies with similar types of military forces, albeit outnumbered. The role of allies, principally NATO, was assumed and taken into account in planning, although the paradox of the issue of planning for a long versus short war in a nuclear world remained unresolved.

Mobilization, as in World War II, was likewise assumed if the Cold War went hot while, at the same time, it was hoped that any war might be ended quickly. The largely World War II defense, industrial, and basing structure was retained along with the intent to rely on our technological superiority to offset numerical or geographical liabilities.

It was not by accident that this Cold War concept of defense through mobilization was similar to the strategy that won the Second World War and the literal ability of ultimately overwhelming the enemy using the massive application of force, technology, and associated firepower. Two decades later, Vietnam exposed the frailty of this approach of dependence on massive application of firepower especially when political limits were placed on applying that firepower.

Currently, *Desert Storm* and the liberation of Kuwait in 1991 have been taken as the examples that confirm the validity of the doctrine of overwhelming or decisive force and of ensuring that both strategic objectives and tactical methods were in congruence. We argue that now is the time to re-examine these premises of reliance on overwhelming or decisive force as currently employed and deployed in the force structure if only as a prudent check.

Beyond prudence, however, it is clear that without a major threat to generate consensus and to rally the country around defense and defense spending, the military posture of the United States will erode as the defense budget is cut. Hence, relying in the future on what is currently seen to be as sufficient force to be "decisive" could easily prove unachievable and the results problematic or worse for U.S. policy.

The absence of a direct and daunting external security threat is, of course, a most obvious aspect of the difficulty in defining the future defense posture of the nation. The United States has long resisted maintaining a large standing military and the Cold War years could prove an aberration to that history. Extending this historical observation of small standing forces, it is clear that there is no adversary on the horizon even remotely approaching the military power of the former USSR. While we might conjure up nominal regional contingencies against Korea or Iraq as sensible planning scenarios for establishing the building blocks for force structure, it will prove difficult to sustain the current defense program over the long term without a real threat materializing to rally and coalesce public support. Allocating three percent or less of GDP for defense could easily prove to be a ceiling and not a floor. It should be noted that in Europe, defense spending is closing in on 1 to 2 percent of GDP.

Ironically, as the Department of Defense seeks to come to grips with this new world, the structural limitations and constraints in how we develop systems and procure weapons based on current technological and industrial capacity for producing them will be exacerbated by downward fiscal pressure giving us little room for mistakes and flexibility. Air, land, space, and sea forces are currently limited in the actual numbers and types of systems that are available for purchase and more limited in that there are virtually no new major systems on the horizon. That could change.

The M-1A-1 tank is in production only for foreign sales. Despite the allure of the Arsenal Ship, the

Background and Basis 27

Navy still has only four active classes of warships from which to replace its capability and, for the first time this century since aircraft entered the inventory, is without a new aircraft in development. The Air Force can be placed in similar straits if the F-22 program is deferred or canceled because of rising cost and fiscal constraints. Time will tell what happens to the Joint Strike Fighter. Assumptions about reliance on technology and R&D providing insurance policies for future defense needs may prove ill-advised if and as DOD is forced to cut back and reduce those programs even further. Indeed, over time, commercial R&D could become the main source for procuring software and other systems needed to upgrade today's weapons systems and for so-called "leap-ahead" technologies that may prove elusive to create.

There is also the crucial issue of revising or indeed developing new doctrine and military thought to deal with these changing circumstances. But, without a compelling rationale and with the clear bureaucratic and political pressures of preparing and defending an annual budget, more of the same (or more likely, less of the same) becomes an almost irresistible outcome. While the JCS or OSD or CINCs may have genuine need for jointly packaged forces that are rapidly deployable irrespective of Army, Navy, Marine, or Air Force labels, the services cannot be expected to reverse the years of viewing the world through service-specific arguments and doctrine.

Although the absolute danger has been dramatically reduced with the end of the USSR, it would be the height of folly to assume that there are no risks to the nation nor an absence of evil-doers wishing this nation harm. It would also be shortsighted to expect that potential adversaries are unintelligent and would not rely on superior knowledge of their environment and simplicity to overcome our current military and technical superiority much as the North Vietnamese did. In addition, as technology diffuses around, over, and under borders, our assumptions about guarantees of permanent technological superiority should welcome thoughtful examination.

Lenin asked the question, "what is to be done?" As a start, the United States should act to exploit the several major advantages it possesses. First, we have time. The clarity and danger of future threats is sufficiently removed for us to take a longer view. While we may have deferred adding to the inventory of future systems in development, current systems possess more than enough military capability to get us through this transition period, even if this period were to last for more than a decade. This does not mean we can rest on our oars; if we take advantage of this opportunity, time is on our side. If we squander this opportunity, then we could ultimately find ourselves in trouble.

Second, the combination of American technical know-how, the luxury of the best technically educated and trained society in the world, and the entrepreneurial spirit of our system offers vast potential if we are clever enough to exploit this extraordinary resource.

Third, because of significant changes in law and organization regarding the military, particularly the Goldwater-Nichols Act, and through a willingness to examine alternatives, the Department of Defense has actively sought new ideas and concepts. The enhanced role of the CINCs and the acceptance of jointness are positive illustrations. Yet, for understandable structural and political realities noted above, assuring productive innovation continues will not be automatic. Against these conducive signs, vision, true joint thinking, and tactical advances still are premium commodities to be nourished and encouraged.

In building an alternative intellectual concept, it is useful to rely on successful lessons of the past. For five decades, we have been successful in applying containment and deterrence in the Cold War. When deterrence or diplomacy failed as in Kuwait, then the use of force was inevitable. A first-order issue is how can we augment or improve the use of existing military capability should it be required.

Should force be needed, our proposal calls for establishing a regime of Rapid Dominance throughout the area of strategic as well as operational concern. By Rapid Dominance, we are seeking the capability to dominate, control, and isolate the entire environment in, around, over, and under the objective area as quickly as possible, and with fewer forces than currently envisaged, although direct insertion of forces is an important component depending upon the tactical situation. In many cases, this capacity need not be the traditional firepower solution of only physically destroying an adversary's military capabilities. Our focus is on the Clausewitzian principle of affecting the adversary's will to resist as the first order of business, quickly if not nearly instantaneously. A second goal would be to stop an attack during the first stages. A third goal, should it be achievable, would be to promote a regime of political deterrence that might restrain aggression in the first place.

To accomplish the rendering an adversary incapable of action means neutralizing the ability to command; to provide logistics; to organize society; and to function; as well as to control, regulate and deny the adversary of information, intelligence, and understanding of what is and what is not happening. This means we must control all necessary intelligence and information on our forces—the ultimate form of stealth—and on an adversary's forces as well and then exploit total situational awareness for rapid action.

Regarding the emergence of current military thought and doctrine, as implied earlier, warfare today may be in the early and far less mature stages of a major revolution than is generally assumed. It is understandable that despite major strategic reassessments, current doctrine is still highly influenced by Cold War tactics and strategy and perhaps by the iron grip of the history of conflict since the early 19th century.

Since Napoleon, the conduct of war between major states has been largely dominated by combining industrial might with vast amounts of manpower over time and space. The United States advanced Napoleon's use of industry and mass armies in the Civil War and our planning up to the Cold War tended to follow this same pattern. World War II, of course, exemplified the triumph of this industrial, mobilization, and massive use of force approach.

In the evolution of U.S. military theory, it can be argued that this model combining massive industrial might and manpower finally ended in 1989. Although, by then, technological advances to conventional military capabilities seemed to be approaching the destructive power, or more precisely, the system lethality of nuclear weapons. In other words, modern non-nuclear precision weapons perhaps could produce effects against enemy targets roughly comparable to the military lethality of theater-level nuclear weapons. If this condition proves true, could this new lethality fundamentally change the construct for designing American doctrine and strategy? This question is at the heart of the "precision and battlefield awareness" school of decisive force thinking that believes that this fundamental change is in place.

Since the end of the Cold War and, with it, the end of the need to prepare our forces to fight a more or less equally powerful adversary, the United States military has conducted two post-Cold War crises against lesser adversaries quite differently than it fought the Cold War. In the Panama intervention in 1990 and in Kuwait shortly thereafter, the suggestion of newer and different methods of warfare was present. Perhaps both will turn out to be transition campaigns, where there is much of the old, but also signs of the new. But there are specific pieces of evidence that should command our attention.

Underlying the planning for *Operation Just Cause* in Panama and *Desert Shield/Storm* in Kuwait was the premeditated incorporation of a series of rapid, simultaneous attacks designed to apply decisive force. The aim was to stun, and then rapidly defeat the enemy through a series of carefully orchestrated land, sea, air, and special operating forces strikes that took place nearly simultaneously across a wide battle space and against many military targets. The purpose of these rapid, simultaneous attacks was to produce immediate paralysis of both the national state and its armed forces that would lead to prompt neutralization and capitulation.

In both *Just Cause* and *Desert Storm*, the United States (plus coalition forces in *Desert Storm*) had such overwhelming military capabilities that, in retrospect, the outcome was largely a matter of drafting a cogent and coordinated operation plan based on using the entire system of capabilities, and then executing that plan to produce a decisive victory. The Haitian incursion in 1995 used similar principles of intimidation to eliminate any real fighting. However, in *Desert Storm* unlike Haiti, it took the U.S. and its allies nearly 6 months to deploy over a half million troops before the fighting began.

The recently published JCS Pub 3.0 and the U.S. Army's 525-5 Pamphlet reflect and exploit operational rapidity and simultaneity. Yet, progress in these operational directions may be in danger of faltering if only old Cold War yardsticks are used to make future force investments and to direct studies about future force structure and associated infrastructure. As in any transition period, innovation must be joined by a willingness to experiment. This means the establishment and cultivation of an experimental apparatus to test and evaluate new concepts are matters of importance both to foster innovation and assess its application.

We build on the trends of rapidity and simultaneity and seek to emphasize control and time. Control is necessary to force behavioral change in adversaries to achieve strategic or political ends. Control and then influence come from a range of threats and outcomes, including putting at risk the targets an adversary holds dear, to imposing a hierarchy of Shock and Awe, to affecting will, perception, and understanding. Achieving control may now be theoretically possible in even more compressed or shortened time periods because of the potential superiority of enhanced U.S. military capability and further training and education. To obtain this level of military superiority that can affect the adversary's will and perception, or at least achieve the practical military consequences, a great deal of thought, debate, and experimentation over new concepts will be needed if only to test and validate contemporary doctrine.

If the political objective is to achieve a level of Shock and Awe beyond only temporary paralysis, then further actions must follow. The end point will be to dominate the enemy in such a way as to achieve the desired objectives. From this concept follows the need to shut down either a state or an organized enemy through the rapid and simultaneous application (or threat of application) of land, sea, air, space, and special operating forces against the broadest spectrum of the adversary's power base and center or centers of gravity and against the adversary's will and perception at tactical and strategic levels.

In *Desert Storm,* the objectives were first to evict Iraqi forces from Kuwait and then to restore the legitimate government. From these objectives, more limited strategic and political objectives followed, some for purposes of maintaining coalition solidarity and UN-imposed sanctions. Not occupying Baghdad was one such political limitation. These strategic objectives led to identification of the enemy's centers of gravity as the basis for the application of force to destroy these centers. This planning led to the repeated, rapid, and simultaneous use of massive force with great effect.

One obvious tactical objective was to eliminate Saddam Hussein's command and control. This was accomplished by simultaneous and massive attacks. Once command and control was destroyed, Iraqi forces in the Kuwait Theater of Operations (KTO) would be destroyed as quickly as possible with overwhelming force and with minimum casualties. As General Colin Powell simply stated, "My plan is to cut off Saddam's army and then kill it."

There was no sanctuary for Iraqi forces in the KTO. They were completely vulnerable to unrelenting and devastating attack. Outside the KTO, targeting was more selective, not because the means were unavailable for imposing sufficient damage but because our military objectives were purposely limited. Given the effectiveness of the air campaign and the overwhelming superiority on the ground, coalition land forces required only 4 of the 41 days of the war to defeat and to eject Iraq's forces from Kuwait.

Suppose a *Desert Storm*-type campaign were fought 20 years from now based on a plan that exploited the concept of Rapid Dominance. Further assume that Iraq has improved (and rebuilt) its military and that, in a series of simultaneous and nearly instantaneous actions, our primary objective was still to shut Iraq down, threaten or destroy its leadership, and isolate and destroy its military forces as we did in 1991. However, two decades hence, Rapid Dominance might conceivably achieve this objective in a matter of days (or perhaps hours) and not after the 6 months or the 500,000 troops that were required in 1990 to 1991. Rapid Dominance may even offer the prospect of stopping an invasion in its tracks.

Shutting the country down would entail both the physical destruction of appropriate infrastructure and the shutdown and control of the flow of all vital information and associated commerce so rapidly as to achieve a level of national shock akin to the effect that dropping nuclear weapons on Hiroshima and Nagasaki had on the Japanese. Simultaneously, Iraq's armed forces would be paralyzed with the neutralization or destruction of its capabilities. Deception, disinformation, and misinformation would be applied massively.

This level of simultaneity and Rapid Dominance must also demonstrate to the adversary our endurance and staying power, that is, the capability to dominate over as much time as is necessary less an enemy mistakenly try to wait it out and use time between attacks to recover sufficiently. If the enemy still resisted, then conventional forms of attack would follow resulting in the physical occupation of territory. Control is thus best gained by the demonstrated ability to sustain the stun effects of the initial rapid series of blows long enough to affect the enemy's will and his means to continue. There must be staying power effect on the enemy or they merely absorb the blows, gain in confidence and their ability to resist, and change tactics much as occurred during the WWII bombing campaigns and the air war over North Vietnam.

Achieving these levels of Shock and Awe requires a wide versatility and competence in employing land, sea, air, space, and special operating forces and in investment in technology to produce Rapid Dominance. Different methods for commanding the battle using both hierarchical and non-hierarchical command to control and direct our forces are likely to be required especially given the simultaneous application of capabilities throughout the given battle space by the full spectrum of our forces. To use these combinations of forces will require adjustment of current service doctrine and prescribed roles and functions. Rapid Dominance also means looking to invest in technologies perhaps not fully or currently captured by the Cold War paradigm.

To develop the proper combination of forces and future technology investment for Rapid Dominance, extensive experimentation with this core concept will be required. This experimentation must apply to all levels of military educational institutions; it must be joint; it can be accelerated by availability of recent advances in simulation technology; and it must have operational trials in the field.

To advance this concept, technology and its infrastructure and application are vital. Here, understanding several facts is important. The U.S. today is graduating through its college and universities system approximately 200,000 American and foreign scientists and engineers per year. This is a great national resource. This technology infrastructure is dimensions larger in number and scope than the aggregate of anywhere else in the world. Through appreciation and exploitation of this potential, a U.S. position of pre-eminence in science and technology could be assured for the foreseeable future.

One adjunct of this technology revolution is in the information and information management areas—which, in the U.S., are heavily commercially oriented. Future military application may well be analogous to the impact of the internal combustion engine and wireless radio on land, sea, and air forces in the 1920s and 1930s. The size of this technological lead between ourselves and the rest of the world, especially in the base for new information products and services, should widen further in knowledge and in application. The "Silicon Valley" revolution is likely to continue increasing computer capacity on an almost annual basis. By the year 2005, computing power should be many fold times today's capacity—perhaps ultimately beginning to close in on the ability of humans to handle data flow as well as the ability to condense and synthesize data.

In parallel to advances in computing power will be the ability to transfer information into and out of the hands of individual users. The addition of virtual reality and other technical aids will enhance and potentially quicken individual decision-making ability. Technologies associated with bioscience and bioengineering are likely to be of particular importance in enhancing these capabilities and are also an area of American predominance. Material sciences, software, and communications are all American strengths, and should remain so well into the next century.

A significant element supporting this explosion in applied information and other technologies is the American free enterprise system and its entrepreneurial character. This drive is needed to translate this technology into military hardware. The nature of the U.S. market and its competitive basis reinforce this element. The largest challenges may be to shape and exploit this commercial potential and then to ensure that its enduring advantages become fundamental in the makeup of our military forces. Unlike the defense industrial base required during the Cold War, this new commercial base is neither heavy nor is it a massive industry relying on producing large things. Indeed, its edge has depended on getting "smaller, smarter, and cheaper."

The fundamental technology thrust for channeling this new American industrial base to support Rapid

Dominance must be toward the control and management of everything that is significant to the operations bearing on the particular Area of Interest (AOI). And we mean everything! Control of the environment is far broader than only the objective of achieving dominant battlefield awareness. Control means the ability to change, to a greater or lesser degree, the "signatures" of all of the combat forces engaged in the AOI. With this concept, the operational frameworks in applying force across the entire spectrum of platforms (satellites, aircraft, land vehicles, ships) can be measured (and controlled) from many minus decibels of cross section, to many plus decibels; communications can be entirely covert, i.e., many dB less than the ambient environment, or that approaching "white noise." The location of both the individual and his unit can be measured in real time in meters, if not feet, anywhere in the world. Through virtual reality, movement in three-dimensional grids over hundreds of square kilometers, offer precise location and movement control, both during day and night in conditions of unprecedented confidence. This occurs in real time. Denying or deceiving the adversary, including real-time manipulation of senses and inputs, is part of this control.

A Rapid Dominance-configured force would enter an AOI and immediately control the operational/ environmental signatures both individually and in the aggregate. As needed, line and non-line-of-sight weapons of near pin-point accuracy would be delivered across the entire area of operation. Stealthy UAVs and mobile robotics systems, together with decoys, would be deployed in large numbers for surveillance, targeting, strike, and deception and would produce their own impact of electronic Shock and Awe on the enemy. This application of force can be done as rapidly as political and strategic conditions demand.

The effects mean literally "turning on and off" the "lights" that enable any potential aggressor to see or appreciate the conditions and events concerning his forces and, ultimately, his society. What is radically different in Rapid Dominance is the comprehensive system assemblage and integration of many evolving and even revolutionary technical advances in dominant battlefield awareness squared—materials application, sensor and signature control, computer and bioengineering applied to massive amounts of data, enable weapon application with simultaneity, precision, and lethality that to date have not been applied as a total system. Deception, disinformation, and misinformation will become major elements of this systemic approach.

The R&D reality is that technology advances will likely come from the commercial world as the DOD base continues to shrink. It is clear that in certain areas, DOD must remain involved where there is no private R&D or to fill gaps in R&D. Warships, fighter aircraft, tanks, and missile defense are examples. However, advances in commercial technology in the Information Age are unlikely to be matched by DOD.

Of equal importance is how we train, organize, and educate our combat officers and key enlisted personnel. Command must be geared to achieving the best of the best—not the best among the good. Assimilating in real time the vast amount of information and putting information to use will no doubt lead to major changes in the composition, competence, and authority of (even and especially) individual military unit commanders perhaps to the squad or private soldier level.

Of course, even with the most perfect information, an unqualified, inexperienced, or unprepared military commander may not win except with extraordinary luck or an incompetent foe. And, we repeat that there are cases where NO military force may be able to succeed if the objectives are unobtainable. The match of the entrepreneurial individual with the potential of the technology base is key. Optimizing and integrating all elements into a total system is a certain way to exploit the opportunity that we can perceive becoming more visible in the coming years.

Shock and Awe

The basis for Rapid Dominance rests in the ability to affect the will, perception, and understanding of the adversary through imposing sufficient Shock and Awe to achieve the necessary political, strategic, and operational goals of the conflict or crisis that led to the use of force. War, of course, in the broadest sense has been characterized by Clausewitz to include substantial elements of "fog, friction, and fear." In the Clausewitzian view, "shock and awe" were necessary effects arising from application of military power and were aimed at destroying the will of an adversary to resist. Earlier and similar observations had been made by the great Chinese military writer Sun Tzu around 500 B.C. Sun Tzu observed that disarming an adversary before battle was joined was the most effective outcome a commander could achieve. Sun Tzu was well aware of the crucial importance of achieving Shock and Awe prior to, during, and in ending battle. He also observed that "war is deception," implying that Shock and Awe were greatly leveraged through clever, if not brilliant, employment of force.

In Rapid Dominance, the aim of affecting the adversary's will, understanding, and perception through achieving Shock and Awe is multifaceted. To identify and present these facets, we need first to examine the different aspects of and mechanisms by which Shock and Awe affect an adversary. One recalls from old photographs and movie or television screens, the comatose and glazed expressions of survivors of the great bombardments of World War I and the attendant horrors and death of trench warfare. These images and expressions of shock transcend race, culture, and history. Indeed, TV coverage of *Desert Storm* vividly portrayed Iraqi soldiers registering these effects of battlefield Shock and Awe.

In our excursion, we seek to determine whether and how Shock and Awe can become sufficiently intimidating and compelling factors to force or otherwise convince an adversary to accept our will in the Clausewitzian sense, such that the strategic aims and military objectives of the campaign will achieve a political end. Then, Shock and Awe are linked to the four core characteristics that define Rapid Dominance: knowledge, rapidity, brilliance, and control.

The first step in this process is to establish a hierarchy of different types, models, and examples of Shock and Awe in order to identify the principal mechanisms, aims, and aspects that differentiate each model as unique or important. At this stage, historical examples are offered. However, in subsequent stages, a task will be to identify current and future examples to show the effects of Shock and Awe. From this identification, the next step in this methodology is to develop alternative mission capability packages consisting of a concept of operations doctrine, tactics, force structure, organizations, and systems to analyze and determine how best each form or variant of Shock and Awe might be achieved. To repeat, intimidation and compliance are the outputs we seek to obtain by the threat of use or by the actual application of our alternative force package. Then the mission capability package is examined in conditions of both MRCs and OOTW.

For discussion purposes, nine examples representing differing historical types, variants, and characteristics of Shock and Awe have been derived. These examples are not exclusive categories and overlap exists between and among them. The first example is "Overwhelming Force," the doctrine and concept shaping today's American force structure. The aims of this doctrine are to apply massive or overwhelming force as quickly as possible on an adversary in order to disarm, incapacitate, or render the enemy militarily impotent with as few casualties and losses to ourselves and to non-combatants as possible. The superiority of American forces, technically and operationally, is crucial to successful application.

There are several major criticisms and potential weaknesses of this approach. The first is its obvious reliance on large numbers of highly capable (and expensive) platforms such as the M-1 tank, F-14,15, and 18 aircraft and CVN/DDG-51/SSN-688 ships designed principally to be used jointly or individually to destroy and attrite other forces and supporting capability. In other words, this example has principally been derived from force-on-forces attrition relationships even though command and control, logistical, and supporting forces cannot be disaggregated from this doctrine.

The other major shortcoming of a force-on-force or a platform-on-platform attrition basis is that with declining numbers of worthy and well enough equipped adversaries against whom to apply this doctrine, justifying it to a questioning Congress and public will prove more difficult. While it is clear that "system of systems" and other alternative military concepts are under consideration, for the time being, these have not replaced the current platform and force-on-force attrition orientation. It should be noted, there will be no doctrinal alternatives unless ample effort is made to provide a comprehensive and detailed examination of possible alternatives.

Second, this approach is based on ultimately projecting large amounts of force. This requires significant logistical lift and the time to transport the necessary forces. Rapidity may not always follow, especially when it is necessary to deliver large quantities of decisive force to remote or distant regions. Third, the costs of maintaining a sufficiently decisive force may outstrip the money provided to pay for the numbers of highly capable forces needed. Finally, at a time when the commercial marketplace is increasing the performance of its products while also lowering price and cycle time to field newer generations systems, the opposite trends are still endemic in the defense sector. This will compound the tension between quality and quantity already cited. None of these shortcomings is necessarily fatal. However, none should be dismissed without fuller understanding.

Certainly, Rapid Dominance seeks to achieve certain objectives that are similar to those of current doctrine. A major distinction is that Rapid Dominance envisages a wider application of force across a broader spectrum of leverage points to impose Shock and Awe. This breadth should lead to a more comprehensive and integrated interaction among all the specific components and units that produce aggregate military capability and must include training and education, as well as new ways to exploit our technical and industrial capacity. It is possible that in these resource, technical, and commercial industrial areas that Rapid Dominance may provide particular utility that otherwise may constrain the effectiveness of Decisive Force.

The second example is "Hiroshima and Nagasaki" noted earlier. The intent here is to impose a regime of Shock and Awe through delivery of instant, nearly incomprehensible levels of massive destruction directed at influencing society writ large, meaning its leadership and public, rather than targeting directly against military or strategic objectives even with relatively few numbers or systems. The employment of this capability against society and its values, called "counter-value" in the nuclear deterrent jargon, is massively destructive strikes directly at the public will of the adversary to resist and, ideally or theoretically, would instantly or quickly incapacitate that will over the space of a few hours or days.

The major flaws and shortcomings are severalfold and rest in determining whether this magnitude and speed of destruction can actually be achieved using non-nuclear systems to render an adversary impotent; to destroy quickly the will to resist within acceptable and probably unachievably low levels of societal destruction; and whether a political decision would be taken in any case to use this type of capability given the magnitude of the consequences and the risk of failure.

It can be argued that in the bombing campaign of *Desert Storm*, similar objectives were envisioned. The differences between this example and *Desert Storm* are through the totality of a society that would be affected by a massive and indiscriminate regime of destruction and the speed of imposing those strikes as occurred to those Japanese cities. This example of shock, awe, and intimidation rests on the proposition that such effects must occur in very short periods of time.

The next example is "Massive Bombardment." This category of Shock and Awe applies massive and, perhaps today, relatively precise destructive power largely against military targets and related sectors over time. It is unlikely to produce an immediate effect on the will of the adversary to resist. In a sense, this is an endurance contest in which the enemy is finally broken through exhaustion. However, it is the cumulative effect of this application of destruction power that will ultimately impose sufficient Shock and Awe, as well as perhaps destroy the physical means to resist, that an adversary will be forced to accept whatever terms may be imposed. As noted, trench warfare of the First World War, the strategic bombing campaign in Europe of the Second World War (which was not effective in this regard), and related B-52 raids in Vietnam and especially over the New Year period of 1972-73, illustrate the application of massive bombardment.

Massive Bombardment, directed at largely military-strategic targets, is indeed an aspect of applying "Overwhelming Force," even though political constraints make this example most unlikely to be repeated in the future. There is also the option of applying massive destruction against purely civilian or "counter-value" targets such as the firebombing of Tokyo in World War II when unconditionality marks the terms of surrender. It is the cumulative impact of destruction on the endurance and capacity of the adversary that ultimately affects the will to resist that is the central foundation of this example.

The shortcoming with this example is clear, and rests in the question of political feasibility and acceptability, and what circumstances would be necessary to dictate and permit use of massive bombardment. Outright invasion and aggression such as Iraq's attack against Kuwait could clearly qualify as reasons to justify using this level of Shock and Awe. However, as with Overwhelming Force, this response is not time-sensitive and would require massive application of force for some duration as well as political support.

Fourth is the "Blitzkreig" example. In real Blitzkreig, Shock and Awe were not achieved through the massive application of firepower across a broad front nor through the delivery of massive levels of force. Instead, the intent was to apply precise, surgical amounts of tightly focused force to achieve maximum leverage but with total economies of scale. The German Wehrmacht's Blitzkreig was not a massive attack across a very broad front, although the opponent may have been deceived into believing that. Instead, the enemy's line was probed in multiple locations and, wherever it could be most easily penetrated, attack was concentrated in a narrow salient. The image is that of the shaped charge, penetrating through a relatively tiny hole in a tank's armor and then exploding outwardly to achieve a maximum cone of damage against the unarmored or less protected innards.

To the degree that this example of achieving Shock and Awe is directed against military targets, it requires skill if not brilliance in execution, or nearly total incompetence in the adversary. The adversary, finding front lines broken and the rear vulnerable, panics, surrenders, or both. Hitler's campaign in France and Holland and the seizure of the Dutch forts and the occupation of Crete in 1940 are obvious illustrations. The use of Special Operations forces in significant numbers is an adjunct to imposing this level of Shock and Awe.

Desert Storm could have been a classic Blitzkreig maneuver if the attack were mounted without the long preparatory bombardment and was concentrated in a single sector—either the "left hook" or the Marine attack "up the middle," and with total surprise. The major differences between the operation in Kuwait and Germany's capture of France in 1940 were that the allies in Saudi Arabia had complete military and technical superiority unlike the Germans and that, once under attack, Iraq's front line collapsed virtually everywhere, giving the coalition license to pick and choose the points for penetration and then dominate the battle with fire and maneuver. The lesson for future adversaries about the Blitzkreig example and the United States is that they will face in us an opponent able to employ technically superior forces with brilliance, speed, and vast leverage in achieving Shock and Awe through the precise application of force.

It must also be noted that there are certainly situations such as guerilla war where this or most means of employing force to obtain Shock and Awe may simply prove inapplicable. For example, the German Blitzkreig would have performed with the greatest difficulty in the Vietnam War, where enemy forces had relatively few lines to be penetrated or selectively savaged by this type of warfare.

The shortcomings of Blitzkrieg ironically rest in its strengths. Can brilliance and superiority be maintained? Is there a flexible enough infrastructure to ensure training to that standard, and can the supporting industrial base continue to produce at acceptable costs the systems to maintain this operational and technical superiority? Rapid Dominance requires a positive answer to these questions, at least theoretically.

The fifth example is named after the Chinese philosopher-warrior, Sun Tzu. The "Sun Tzu" example is based on selective, instant decapitation of military or societal targets to achieve Shock and Awe. This discrete or precise nature of applying force differentiates this from Hiroshima and Massive Destruction examples. Sun Tzu was brought before Ho Lu, the King of Wu, who had read all of Sun Tzu's thirteen chapters on war and proposed a test of Sun's military skills. Ho asked if the rules applied to women. When the answer was yes, the king challenged Sun Tzu to turn the royal concubines into a marching troop. The concubines merely laughed at Sun Tzu until he had the head cut off the head concubine. The ladies still could not bring themselves to take the master's orders seriously. So, Sun Tzu had the head cut off a second concubine. From that point on, so the story goes, the ladies learned to march with the precision of a drill team.

The objectives of this example are to achieve Shock and Awe and hence compliance or capitulation through very selective, utterly brutal and ruthless, and rapid application of force to intimidate. The fundamental values or lives are the principal targets and the aim is to convince the majority that resistance is futile by targeting and harming the few. Both society and the military are the targets. In a sense, Sun Tzu attempts to achieve Hiroshima levels of Shock and Awe but through far more selective and informed targeting. Decapitation is merely one instrument. This model can easily fall outside the cultural heritage and values of the U.S. for it to be useful without major refinement. Shutting down an adversary's ability to "see" or to communicate is another variant but without many historical examples to show useful wartime applications.

A subset of the Sun Tzu example is the view that war is deception. In this subset, the attempt is to deceive the enemy into what we wish the enemy to perceive and thereby trick, cajole, induce, or force the adversary. The thrust or target is the perception, understanding, and knowledge of the adversary. In some ways, the ancient Trojan Horse is an early example of deception. However, as we will see, the deception model may have new foundations in the technological innovations that are occurring and in our ability to control the environment.

The shortcomings with Sun Tzu are similar to those of the Massive Destruction and the Blitzkreig examples. It is questionable that a decision to employ American force this ruthlessly in quasi- or real assassination will ever be made by the U.S.

Further, the standard to maintain the ability to perform these missions is high and dependent on both resources and on supporting intelligence, especially human intelligence—not an American strong point.

Britain's Special Air Service provides the SAS example and is distinct from the Blitzkreig or Sun Tzu categories because it focuses on depriving an adversary of its senses in order to impose Shock and Awe. The image here is the hostage rescue team employing stun grenades to incapacitate an adversary, but on a far larger scale. The stun grenade produces blinding light and deafening noise. The result shocks and confuses the adversary and makes him senseless. The aim in this example of achieving Shock and Awe is to produce so much light and sound or the converse, to deprive the adversary of all senses, and therefore to disable and to disarm. Without senses, the adversary becomes impotent and entirely vulnerable.

A huge "battlefield" stun grenade that encompasses large areas is a dramatic if unachievable illustration. Perhaps a high altitude nuclear detonation that blacks out virtually all electronic and electrical equipment better describes the intended effect regardless of likelihood of use. Depriving the enemy, in specific areas, of the ability to communicate, observe, and to interact is a more reasonable and perhaps more achievable variant. This deprival of senses, including all electronics and substitution of false signals or data to create this feeling of impotence, is another variant. Above all, Shock and Awe are imposed instantly and the mechanism or target is deprivation of the senses.

The shortcomings of the SAS approach mirror in part shortcomings of other approaches. Technological solutions are crucial but may not be conceivable outside the EMP effects of nuclear weapons. Intelligence is clearly vital. Without precise knowledge of who and what are to be stunned, this example will not work.

The sixth example of applying Shock and Awe is the "Haitian" example (or to the purist, the Potemkin Village example). It is based on imposing Shock and Awe through a show of force and indeed through deception, misinformation, and disinformation and is different from the U.S. intervention in Haiti in 1995. In the early 1800s, native Haitians were seeking to extricate their country from French control. The Haitian leaders staged a martial parade for the visiting French military contingent and marched, reportedly, a hand full of battalions repeatedly in review. The French were deceived into believing that the native forces numbered in the tens of thousands and concluded that French military action was futile and that its forces would be overwhelmed. As a result, the Haitians were able to achieve their freedom without firing a shot.

To be sure, there are points of similarity between the Haitian example and the others. Deception, disinformation, and guile are more crucial in this regime. However, the target or focus is the will and perception of the intended target. Perhaps the Sun Tzu category comes closest to this one except that while Sun Tzu is selective in applying force, it is clear that imposing actual pain and shock are essential ingredients and deception, disinformation, and guile are secondary. Demonstrative uses of force are also important. The issue is how to determine what demonstrations will affect the perceptions of the intended target in line with the overall political aims.

The weakness of this form of Shock and Awe is its major dependency on intelligence. One must be certain that the will and perceptions of the adversary can be manipulated. The classic misfire is the adversary who is not impressed and, instead, is further provoked to action by the unintended actions of the aggressor. Saddam Hussein and the Iraqis' invasion of Kuwait demonstrate when this Potemkin Village model can backfire. Saddam simply let his bluff be called.

The next example is that of "The Roman Legions." Achieving Shock and Awe rests in the ability to deter and overpower an adversary through the adversary's perception and fear of his vulnerability and our own invincibility, even though applying ultimate retribution could take a considerable period of time. The target set encompasses both military and societal values. In occupying a vast empire stretching from the Atlantic to the Red Sea, Rome could deploy relatively small number of forces to secure each of these territories. In the first place, Roman forces were far superior to native forces individually and collectively. In the second place, if an untoward act occurred, the perpetrator could rest assured that Roman vengeance ultimately would take place. This was similar to British "Gunboat Diplomacy" of the nineteenth century when the British fleet would return to the scene of any crime against the crown and extract its retribution through the wholesale destruction of offending villages.

There were several vital factors in Rome's ability to achieve Shock and Awe. The invincibility of its Legions, or the perception of that prowess, and the inevitability of retribution were among the most significant factors. In other words, reprisals and the use of force to exact a severe punishment, as well as the certainty that this sword of Damocles would descend, were essential ingredients. The distinction between this category and the others is the ex post facto nature of achieving Shock and Awe. In the other categories, there is the need for seizing the initiative and applying contemporaneous force to achieve Shock and Awe. With the Roman example, the Shock and Awe have already been achieved. It is the breakdown of this regime or the rise of new and as yet unbowed adversaries that leads to the reactive use of force.

The major shortcoming is the assumption of the inevitability of reprisals and the capacity to take punitive action. That is not and may not always be the case with the United States, although we can attempt to make others believe it will be. The takeover of the Embassy in Tehran by dissident "students" in 1979 and American impotence in the aftermath are suggestive of the shortcoming. That aside, the example or perception of the invincibility of American military power is not a bad one to embellish.

The next category for achieving Shock and Awe is termed the Decay and Default model and is based on the imposition of societal breakdown over a lengthy period but without the application of massive destruction. This example is obviously not rapid but cumulative. In this example, both military and societal values are targets. Selective and focused force is applied. It is the long-term corrosive effects of the continuing breakdown in the system and society that ultimately compels an adversary to surrender or to accept terms. Shock and Awe are therefore not immediate either in application or in producing the end result. Economic embargoes, long-term policies that harass and aggravate the adversary, and other types of punitive actions that do not threaten the entire society but apply pressure as in the Chinese water torture, a drop at a time, are the mechanisms. Finally, the preoccupation with the decay and disruption of society produces a variant of Shock and Awe in the form of frustration collapsing the will to resist.

The significant weakness of this approach is time duration. In many cases, the time required to impose such a regime of Shock and Awe is unacceptably long or simply cannot be achieved by conventional or politically acceptable means.

The final example is that of "The Royal Canadian Mounted Police," whose unofficial motto was "never send a man where you can send a bullet." The distinction between this example and the others is that this example is even more selective than Sun Tzu and implies that standoff capabilities as opposed to forces in place can achieve the required objectives. There should not be too fine a point, however, in belaboring differences with the other examples in this regard over standoff. A stealthy aircraft bombing unimpededly is not distinct from a cruise missile fired at 1,000 miles regarding the effect of ordnance on target.

A few observations about these examples offer insights on which to test and evaluate means of applying Rapid Dominance. It is clear that the targets in each category include military, civilian, industrial, infrastructure, and societal components of a country or group. In certain cases, time is the crucial consideration in imposing Shock and Awe and in most of the examples, emphasis is on a rapid or sudden imposition of Shock and Awe. However, in several examples, the effects of Shock and Awe must be and are cumulative. They are either achieved over time or achieved through earlier conditioning and experiences. Not all of these categories are dependent on technology or on new technological breakthroughs. What is relatively new or different is the extent to which brilliance and competence in using force, in understanding where an adversary's weak points lie and in executing military operations with deftness, are vital. While this recognition is not new, emphasis is crucial on exploiting brilliance and therefore on the presumption that brilliance may be taught or institutionalized and is not a function only of gifted individuals.

There is also a key distinction between selective or precise and massive application of force. Technology, in the form of "zero CEP" weapons, may provide the seemingly contradictory capability of systems that are both precise and have the net consequence of imposing massive disruption, destruction, or damage. This damage goes beyond the loss of power grids and other easily identifiable industrial targeting sets. Loss of all communications can have a massively destructive impact even though physical destruction can be relatively limited.

In some of the examples, the objective is to apply brutal levels of power and force to achieve Shock and Awe. In the attempt to keep war "immaculate," at least in limiting collateral damage, one point should not be forgotten. Above all, war is a nasty business or, as Sherman put it, "war is hell." While there are surely humanitarian considerations that cannot or should not be ignored, the ability to Shock and Awe ultimately rests in the ability to frighten, scare, intimidate, and disarm. The Clausewitzian dictum concerning the violent nature of war is dismissed only at our peril.

For a policy maker in the White House or Pentagon and the concerned Member of Congress with responsibility for providing for the common defense, what lessons emerge from these examples and hierarchies? First, there are always broader sets of operational concepts and constructs available for achieving political objectives than may be realized. Not all of these alternatives are necessarily better or feasible. However, the examples suggest that further intellectual and conceptual effort is a worthwhile investment in dealing with national security options in the future.

Second, time becomes an opportunity as well as a constraint in generating new thinking. In many past cases, time was generally viewed as an adversary. We had to race against several clocks to arrive "firstest with the mostest," to prevent an enemy from advancing, or to ensure we had ample forces on station should they be required. Rapid Dominance would alleviate many of these constraints as we would have the capacity to deploy effective forces far more quickly. Therefore, in this case, we can view time as an ally. The political issue rests in longstanding arguments to limit the President from having the capacity to deploy or use force quickly, thereby involving the nation without conferring with full consultation with Congress. While this is an obvious point, it should not eliminate alternative types of force packages derived from Rapid Dominance from full consideration and experimentation. Indeed, our experience with nuclear weapons and emergency release procedures shows that delegating instant presidential authority can be handled responsibly.

Responding to the precise, rapid, and massive criteria of several models, it is clear that one capability not presently in the arsenal is a "zero-CEP" weapon, meaning one that is precise and timely. It is also clear that, while deception, guile, and brilliance are important attributes in war, there are no guarantees that they can be institutionalized in any military force.

Another capability that Rapid Dominance would stress relates to the Sun Tzu example. Suppose there are "EMP-like" or High Powered Microwave (HPM) systems that can be fielded and provide broad ability to incapacitate even a relatively primitive society. In using these weapons, the nerve centers of that society would be attacked rather than using this illustrative system to achieve hard target kill because there were few hard targets. To be sure, HPM and EMP-like systems have been and are being carefully researched.

Finally, to return to the idea that deception, disinformation, and misinformation are crucial aspects of waging war, Rapid Dominance would seek to achieve several further capabilities. By using complete signature management, larger formations could be made to look like smaller and smaller formations made to seem larger. At sea, carrier battle groups could be disguised and smaller warships could be made to appear as large formations. This signature management would apply across the entire spectrum of the senses and not just radar or electronic ranges.

Indeed, gaining the ability to regulate what information and intelligence are both available and not available to the adversary is a key aim. This is more than denial or deception. It is control in the fullest sense of the word.

The next step is to match the four significant characteristics that define Rapid Dominance—knowledge, rapidity, brilliance, and control—with Shock and Awe against achievable military objectives in order to derive suitable strategies and doctrines, configure forces and force packages accordingly, and determine those integrated systems and innovative uses of technologies and capabilities that will provide the necessary means to achieve these objectives in conditions that include both the MRC and OOTW.

Strategic, Policy, and Operational Application

In assessing the future utility and applicability of Rapid Dominance, it is crucial to consider the political context in which force is likely to be employed. As we enter the next century, the probability is low that an overriding, massive, direct threat posed by a peer-competitor to the U.S. will emerge in the near term. Without compelling reasons, public tolerance toward American sacrifice abroad will remain low and may even decrease. This reluctance on the part of Americans to tolerate pain is directly correlated to perceptions of threat to U.S. interests. Without a clear and present danger, the definition of national interest may remain narrow.

Americans have always appreciated rapid and decisive military solutions. But, many challenges or crises in the future are likely to be marginal to U.S. interests and therefore may not be resolvable before American political staying power is exhausted. In this period, political micromanagement and fine tuning are likely to be even more prevalent as administrations respond to public sentiments for minimizing casualties and, without a threat or compelling reason, U.S. involvement.

Future actions and measures may likely reflect "politically correct" alternatives. In 1991, the Gulf War came close to presenting the nearly optimal situation for prosecution to a decisive and irreversible conclusion. Such a course, however, was not politically feasible because it would have shattered the allied coalition while exceeding the authority of the UN mandate. Military operations that impact across a whole population or cause "innocent civilians" to suffer (e.g., some economic sanctions, collateral damage from raids) also are likely to be only politically acceptable in aggravated situations. For example, if economic sanctions cause malnutrition or other health problems or collateral damage from bombing or shelling impacts hospitals, schools, orphanages, or refugee camps, the policy may be the ultimate victim.

The U.S. military is more likely to find itself in a supporting foreign policy role with discrete missions that are only one facet of a larger political context. This context is almost certainly going to expand into militarily grey areas of OOTW, including those impinging on law enforcement and ensuring political stability. Forces may be called upon to deal with or control situations on the margin rather than to achieve total submission or defeat of an opponent. The prevailing political preference is likely to continue to be to try to bound these complex challenges through fine tuning, artificial constructs, and discretely limited tasks, often performed in the midst of internal conflict. Economic sanctions (e.g., Serbia, Iraq), "no fly" zones (e.g., Southern and Northern Iraq and Bosnia), "safe havens" (e.g., Bosnia), humanitarian relief delivered by "all means necessary" (e.g., Somalia, Bosnia), and embassy protection and evacuation (e.g., Liberia in 1991 and again in 1996) are the kinds of OOTW tasks more likely to be assigned by policy makers. Such tasks tend to be inconclusive and of long duration. They also increase vulnerability to terrorist attack such as the bombing of the Kolbah Barracks in Riyadh in June 1996.

Americans prefer not to intervene, especially when the direct threat to the U.S. is ambiguous, tenuous, or difficult to define. Therefore, when intervention is necessary there is likely to be both a political and practical imperative to have allied or international involvement or at least the political cover of the UN, NATO, or appropriate NGOs.

As more states (and sub-national groups) acquire nuclear, chemical, and biological weapons of mass destruction (WMD) capabilities and longer range delivery means, the ability for rogues to inflict pain will increase as will the ability to ratchet up the political risks. WMD can easily complicate our ability to influence positive and constructive behavior of possessors. Because of the threat of retaliation, WMD capabilities may become politically acceptable targets provided collateral damage to civilians is minimized. Preemption may become a more realistic option along the lines of Israel's strikes against Syria's nuclear reactors in 1982. It is, however, a responsible state's worst nightmare to have successfully struck a chemical, biological, or nuclear production facility with precision only to learn the next day that hundreds of civilians have been killed due to the inadvertent release of chemical, biological, or nuclear materials.

There must also be an appropriate political context that justifies the use of preemptive force, as opposed to less destructive or non-lethal types of sanctions (e.g., responses to terrorism in the case of Libya, invasion of Kuwait by Iraq, exports of WMD to a threatening country such as Iran, the North Korean threat to South Korea and Japan).

The U.S. will, nevertheless, need to maintain the capability to deter and defeat both strategic and other direct threats to its vital interests, preferably on a decisive basis. In an unsettled, less structured, and volatile world, the ability to use force with precision, effectiveness, impunity, and, when needed, rapidity, will still be a powerful influence on cooperation, stability, and, where relevant, submission.

Imposing Rapid Dominance on a nation, group, or situation, if achievable, will be a highly desirable and relevant asset in this turbulent period. Bosnia offers an example. At the outset of the breakup of Yugoslavia, if we had had this type of capability, without potentially high costs, to counter effectively the widely predicted invasion of Bosnia, the U.S. strategy for dealing with that tangled and messy situation might have been much different. Thousands of lives might have been spared. In other grey or marginal situations Rapid Dominance could make the difference between a politically acceptable response or inadequate action with consequences similar to what happened in Bosnia.

In considering how Rapid Dominance might apply and might be used, it is first important to know what it is that we want to achieve with military force. We need to consider whether the application of force will allow us to influence and control an adversary's will or merely exacerbate a bad situation. Therefore, it is essential to know what is of value to that adversary. An objective, realistic, and in-depth situational grasp will be essential to such an understanding. For example, disarming or destroying may produce unintended consequences. For a conventional foe that values its military and depends on technology, Rapid Dominance should be particularly effective and persuasive. In the case of less developed nations, however, the opportunity for exercising influence in this way and against military formations may be considerably less and must be carefully assessed.

As noted, in cases of marginal direct threats to U.S. security, the cost in casualties needs to be low. To be effective, we must take away an opponent's ability to make it cost us in terms of casualty levels we consider intolerable. In applying Rapid Dominance we also must be defending something which is of value to us. The lower the value in terms of our national interests, the lower the price we are likely to be willing to pay.

In MRC situations, we need to have the capability to defeat, destroy, or incapacitate an opponent. On the other hand, in OOTW, other non-military factors are likely to be involved and goals made more limited. For example, it may be necessary to intimidate or capture the leadership in order to restore order or reverse an action, or it may simply be necessary to anticipate, prevent, and counter opposition to conduct of a more limited mission (e.g., feeding the starving or protecting innocent people from genocide).

In U.S. planning for OOTW, it is a virtual given that risk will be minimized and there will be a discrete and proportional use of force with minimal collateral damage. This means that there must be a belief that a mission can be accomplished and is worth the resources necessary to do so. Before initiating action in these often confusing situations, objectives must be clearly established and, once engaged, there should be a willingness to persevere through the inevitable rough patches.

Whether in an MRC or in OOTW, we first will need to know what we want to achieve with Rapid Dominance. This is a task for political leadership which is informed with military advice concerning what is feasible, what is not, and what is uncertain. The extent of the mission must be clearly defined. Is it to defeat an enemy so it will no longer pose a threat? Do we only need to stop an adversary from carrying out a particular act? Must we control a situation entirely or only sufficiently to be able to carry out a specific mission? Can we really affect the adversary's will?

Recent events give us examples of outcomes likely to be relevant in the future. MRCs call for the full spectrum of outcomes—from reversing military action (e.g., the invasion of Kuwait); to establishing a government more acceptable to the U.S. and the world, probably using military coercion (Haiti, Panama); to eliminating a threat to the U.S. or its allies. We may want to persuade an adversary to cease an aggression or act of interference or otherwise change behavior we cannot accept or tolerate. Political expectations in MRCs are for the effective use of force and for rapid success or at least steady progress. Casualties should be moderate or at least acceptable, with the threshold of American pain dependent on the directness of the threat to U.S. interests and with the degree of compellance appropriate to the political rationale.

OOTW present a different set of challenges. These challenges are likely to require discrete dominance of specific circumstances rather than total dominance. The general tasks may include a wide variety of requirements. For example, it may be necessary to try to prevent or stop genocide (e.g., Rwanda) and ethnic cleansing (e.g., Bosnia). The task may be to cooperate with a humanitarian relief effort (e.g., prevention of starvation in Somalia or Bosnia). The goal of employing force may be free and fair elections (e.g., Cambodia, Bosnia). The requirement could be to destroy a limited objective (e.g., an above-ground or underground chemical weapons plant or documented nuclear weapons facilities developed by hostile or unfriendly states).

Other tasks could simply be to preserve international rights (e.g., protecting the neutral shipping of the western oil flow in the Gulf during the Iran-Iraq war). A more testing challenge might be to accomplish a limited political goal (e.g., gesture to deal with Israeli incursion in Lebanon in 1982). We undoubtedly will face the future requirement to reverse a potential threat to Americans or to a region of importance with a limited military action (e.g., in Grenada in 1983 or the Mayaguez rescue in Cambodia in 1975). Discrete moves to bolster preventive diplomacy and/or overt measures to demonstrate preparedness to assist (e.g., forces sent to Sudan to support Chad under threat of invasion from Libya and recent Navy operations in the Taiwan Strait) will still be relevant.

Countering terrorism also will be part of a continuing agenda (hostage rescue—e.g., Iran, Lebanon; hijacking—e.g., *Achille Lauro*; deterrent to further moves—e.g., the Higgins operation, Libyan raids, missile attack on Iraq after the threat to former President Bush). We may also need to interdict weapons, terrorists, or other discrete cargoes moving between nations (e.g., North Korean missile shipments to Iran, Iranian and Libyan arms exchanges).

Economic sanctions are likely to continue to be a preferable political alternative or a necessary political prelude to an offensive military step (e.g., implemented as the first step in actions to counter Libyan-sponsored terrorism; tried first as an alternative to war with Iraq; used ineffectively against the Serbs to try to convince them not to continue to support Bosnian Serb aggression; and tried with Haiti as an unsuccessful alternative to occupation). Our past experience has been that we seldom have had decisive or immediate results from these economic measures, sanctions, and embargoes. Considerable time is required to have impact and we have not been particularly efficient in controlling the leakage and spillover in these situations. Sanctions almost always require full international cooperation which cannot be assumed or guaranteed. In Bosnia, of course, some portions of the arms embargo were deliberately allowed to be permeable and the U.S. turned a blind eye to Iran's support of the Bosnians.

Past experience also has taught us some relevant lessons about the potential of Shock and Awe. Improvements in the capabilities enhancing these outcomes could make a decisive difference in dealing with future challenges. History also cautions us as well that there will be restraints in employing Rapid Dominance and that there are fundamental differences in MRC and OOTW applications.

Shock and Awe, when properly applied, have been very effective in the past. They will be effective in the future, even when applied in limited ways that

Strategic, Policy, and Operational Application 79

do not reflect the more encompassing impact envisioned by Rapid Dominance. There are many examples of how a very limited application of force made a significant difference through the mechanisms of Shock and Awe. Experiences, including successes and failures, illustrate some of the potential of Rapid Dominance if implemented effectively.

The Vietnam War provides certain lessons. When B-52 strikes, which made the ground rumble, were added to the equation during the Christmas 1972 bombing of Hanoi, dragging negotiations with the North Vietnamese on a peace agreement moved swiftly to an acceptable conclusion. Daily reports following the controversial B-52 "carpet" bombing raids in Cambodia talked of North Vietnamese/Vietcong soldiers wandering around in a daze due to shock and concussion. Both B-52s and naval gunfire, especially from 16 inch guns of a battleship, had a similar impact on invading North Vietnamese troop concentrations. The mining of Haiphong Harbor, although initiated late in the war, was equally effective in immediately stopping shipping in and out of North Vietnam.

When President Nixon wanted to deal with the perplexing problems of our POWs and failing domestic morale, as well as take away substantial political leverage from the North Vietnamese, he directed the raid to rescue prisoners jailed just outside Hanoi. The raid itself was well executed. American forces reached and searched the prison and returned safely. But no Americans were freed because a last minute transfer of the POWs from the prison had not been detected. If there had been prisoners still there to be rescued, the operation would have been a highly dramatic and influential event. The point is that accurate and timely intelligence remains crucial.

There seems to be little doubt that the combined F-111 and naval air strike against Libya in 1986 in response to the discotheque terrorist attack in Germany gave Gadhafi pause. The perception that he personally might be targeted appeared to get Gadhafi's attention.

When our troops were having difficulty dislodging Grenadian soldiers from their main fortress, Marine tanks were sailed around the island to confront them. At the sight of tank guns, the seemingly stubborn occupants surrendered almost immediately without a fight.

The cease fire in the bloody Iran-Iraq war was quick to follow after the commencement of daily Iraqi long-range rocket bombardments of Tehran that amounted to a reign of terror. Given that both sides were exhausted at that point, a show of force could have been convincing. Strong U.S. action in response to Iran's mining of neutral waters may also have had a sobering effect on the mullahs. Not only were Iran's vulnerable oil-producing platforms in the Gulf boarded and destroyed with impunity by the U.S., but Iranian naval forces that had come out to challenge the U.S. Navy were destroyed. Iraq's reign of terror, and the strong American message to Iran, possibly helped end the war.

In our troublesome stay in Somalia, AC-130 gunships earned immediate respect from potential troublemakers with their ability to see wide areas night or day, remain on station for hours as night patrols, and strike with precision and relative impunity. The methodical drone of AC-130s circling in the air was enough to restore some order, although a few civilians found the noise unsettling. In another situation, the aftermath of systematic UN efforts to destroy faction leader Mohamed Aideed's illegal arms facilities generated an unexpected reaction from other warlords, including those colluding with him, which was to volunteer to hand over their own weapons storage areas. For a fleeting moment, Shock and Awe created an important opportunity.

During the many vagaries of the Bosnia tragedy, it would appear that when NATO accurately delivered potent doses of air power, rather than occasional pin pricks, the Serbs seemed finally to understand that an appearance of cooperation rather than defiance was in their interest. This NATO message in the form of air power, of course, was strengthened by the effectiveness of the accompanying Croatian/Muslim counter-offensive and the fatigue of Bosnian Serb fighters. Sustaining the shock effect with forces on the ground was a necessary combination to gain the staying power effect to change the will of the Serbs. It was not accomplished by air alone. Timing remains important.

Past failures also offer examples of how Rapid Dominance might have made a difference in reacting to those difficult situations. Rapid Dominance might have provided a better response to those setbacks or might have offered a more effective alternative that would have avoided the vulnerabilities in those situations in the first place (e.g., Bay of Pigs, Iran embassy rescue in 1980, Lebanon Marine barracks bombing in 1983, response to the Pueblo seizure by North Korea in 1968, and the reaction to the downed helicopters during the Ranger raid in Somalia).

We should also learn from other states who have demonstrated effective application of the characteristics of Rapid Dominance. Israel's rout of Syria's air force and missile defenses in Lebanon's Baaka Valley shows how dramatic success can have political spillover. On the other hand, Japan's surprise attack on Pearl Harbor produced the reverse effects of Shock and Awe and had the unintended consequence of galvanizing the U.S. into action.

Even without a Rapid Dominance capability or when facing a more technologically dependent opponent, it is clear from these examples and many others in recent U.S. experiences that certain improvements in capabilities would provide us with greater flexibility in the future. This is especially true in OOTW situations, which require a multiplicity of effective instruments at our disposal. It is also true that certain operations such as peacekeeping tend to be manpower intensive.

If we are to stay ahead of an adversary and deny things of value to that adversary, dynamic, accurate, and integrated intelligence is essential. Intelligence needs to move to levels unprecedented in scope, timeliness, accuracy, and availability in real time. The Gulf War, despite its success, showed classic limitations in intelligence. Even though we had nearly every intelligence asset designed to deal with the USSR available for use, we were unable to detect the full extent of Iraq's WMD capability; unable to find mobile missile launchers even with a major expenditure of on-scene assets; in some cases, we could only "see" kilometers in front of our advancing forces; and we mistakenly attacked targets we thought were legitimate but had civilians inside. In some instances, only reliable human intelligence may provide the necessary information (for example, in order to understand what is happening in deep underground facilities).

Another important capability we should try to achieve in the future is the ability to intimidate, capture, convince, or significantly influence the perceptions and understanding of individual troublemakers. This need has been demonstrated repeatedly in recent years (e.g., Gadhafi in Libya, Aideed in Somalia, Saddam Hussein in Iraq, Noreiga in Panama). Such a perception is particularly relevant when the problem appears not to be caused by a unified population but by the ambitions of individual leaders who have intimidated or killed off any likely internal opposition. Such a capability requires effective real-time intelligence and a variety of methods for accomplishing the task (from exceptionally precise weapons to effective "snatch" operations).

In a world in which non-lethal sanctions are a political imperative, we will continue to need the ability to shut down all commerce into and out of any country from shipping, air, rail, and roads. We ought to be able to do this in a much more thorough, decisive, and shocking way than we have in the past. The ability to apply pressure or cause acquiescence employing non-lethal means also will be important in some circumstances. Weapons that shock and awe, stun and paralyze, but do not kill in significant numbers may be the only ones that are politically acceptable in the future. This also means that crowd control with minimum violence may be needed. In certain circumstances, the costs of having to resort to lethal force may be too politically expensive in terms of local support as well as support in the U.S. and internationally.

As is already well recognized, we need to be able to shut down key electronic communications to, from, and within a country (or within a specific sub-group or faction). We also need the ability to control radio and television within a country. It is important, however, in all cases, to be able to deny an adversary's ability to communicate and to have our own means of reaching the population with appropriate messages.

In addition to being able to eliminate military capabilities selectively, including weapons systems, overt and covert stockpiles, fuel, WMD, and related logistics, we will need to have the capability selectively to incapacitate, neutralize, or destroy other things considered of great value to opponents. Increased targeting precision will compound effectiveness as well as help to avoid the political pitfalls of using force such as the inevitable, unintended collateral damage that has been the pattern of the past.

More surgical and carefully crafted applications of force, however, will only partially reduce the restraints and limits on utilizing Rapid Dominance in MRCs and OOTW. There are substantial differences in the political constraints likely to be imposed in dealing with MRCs and with OOTW. For example, there is much greater latitude to use dominant force and Shock and Awe in MRCs than in OOTW.

In MRC situations, we are often likely to face conventional powers which are well organized, well equipped, and broadly dependent on technology. Although more powerful, these developed states are also likely to be especially vulnerable to a technologically sophisticated approach such as Rapid Dominance as long as we maintain this military edge and the ability to neutralize their military systems. Even in the most compelling circumstance where a Rapid Dominance force is used, however, support from other nations will be politically desirable.

In most circumstances there will be limits to the targets of value to an adversary which can be destroyed as well as to the numbers and types of weapons that can be employed. For example, the political circumstances in which nuclear weapons could be employed are quite limited. In both MRCs and OOTW, certain actions are politically as well as morally unacceptable except in extreme cases. Such restrictions are likely to apply to targets affecting control of access to food, water, and clean air, and to destruction of religious and cultural centers, even if there is low collateral damage.

In OOTW situations, we are much more vulnerable to criticism of using excessive force, especially if there is civilian or collateral damage. The concept of proportionality is likely to remain an operative principle in U.S. policy and may be taken to extremes, especially if the marginal nature of a situation leads to a marginal and ineffective response. Some people, both military and civilian, even argue that superior technology should not be employed in such situations and that an adversary should be fought on his own terms. While such arguments should be rejected, they nonetheless sometimes have a political influence that must be considered. We should always use technology to minimize our casualties, give us every advantage, reduce collateral damage, and make us look more formidable. At the same time, there needs to be sufficient provocation to warrant destruction or denial. Our actions must always be consistent with our own system of values.

The "rapid" component of Rapid Dominance is one of the most appealing aspects of the concept, both politically and militarily. The ability to take action that is timely and decisive multiplies substantially the chances of ultimate success. Action needs to be taken precisely when it will have greatest impact. Often initial public outrage and political support for action in response to a provocation subsides if a prolonged buildup is necessary in order to prepare to take action.

The ability to react faster than an adversary, to assimilate information and act on it effectively, is also an important advantage. In a NATO region-wide dynamic computer war game a few years ago, it was clear that the simulated enemy was advancing faster than the defensive chain of command could make counter moves. The tradition of sending decisions up the line was simply too slow to cope with the dynamic challenge posed by the adversary. Commanders on scene lacked the authority to respond and adjust to rapidly changing situations. The exercise graphically demonstrated to the country involved the need to institute fundamental command and control streamlining. It also demonstrated the advantages of being able to make local decisions in real time while still effectively coordinating and optimizing the overall effort.

The Navy's "command by negation" concept evolved in the 1980s in order to deal with the rapidity of the air/missile threat and the need to integrate dynamically the offensive and defensive missile, air, sea, and undersea capabilities of a battle group and its joint components (e.g., AWACs). This concept was one way of solving the time problem while keeping the overall commander in the picture. The commander could then intervene and modify actions as necessary to conform to the broader strategy. This type of control was helped by the evolution of electronic links and secure communications and the availability of satellites.

Commanders employing Rapid Dominance will need to orchestrate it using similar principles, while applying greater selective ability to turn on and off a variety of systems, sensors, and devices influencing the whole operational picture. Technology should also give commanders a much better grasp of what is evolving during a battle. Just as the American military of today has made "owning the night" part of its tactical advantage, "owning" the dimension of time will be critical to the success of Rapid Dominance.

In conceptual terms, the following is suggestive of a future force configuration and the design of a mission capability package (MCP) based on Rapid Dominance.

Operational Construct

Rapid Dominance is based on affecting the adversary's will, perception, and knowledge through imposing sufficient Shock and Awe to overcome resistance, allowing us to achieve our aims. Four characteristics are vital: knowledge, rapidity, brilliance, and control of the environment.

Application of all or of selective capabilities within the Rapid Dominance systems of systems will then decisively direct the application of military/defense resources and produce the requisite outcome. Rapid Dominance envisages the execution of specific actions in real or near real time to counter actions or intentions deemed detrimental to U.S. interests. On the high end of conflict, Rapid Dominance would introduce a reaction of Shock and Awe in areas of highest value to the threatening individual, group, or state. In many cases, prior understanding of the power of Rapid Dominance would act as a deterrent to the objectionable action. When used, Rapid Dominance would ensure favorable early resolution of issues with minimal loss of lives and collateral damage. The concept theoretically should be able to impact adversarial situations that apply across the board to high, mid, low, no, or minimal technology threats.

Strategic, Policy, and Operational Application 91

Rapid Dominance expands the art of joint combined arms war fighting capabilities to a new level. Rapid Dominance requires a sophisticated, interconnected, and interoperable grid of netted intelligence, surveillance, reconnaissance, communications systems, data analysis, and real-time deliverable actionable information to the shooter. This network must provide total situational awareness and supporting nodal analysis that enables U.S. forces to act inside the adversary's decision loop in a manner that on the high end produces Shock and Awe among the threat parties. Properly detailed nodal analysis of this knowledge grid will enable the shutting down of specific functions or all essential functions near simultaneously. This will often times be netted pieces of data where the sum of the parts gives the answer and the battlefield advantage to the force possessing this rapidly netted information.

The "rapid" part of the equation becomes the ability to get real-time actionable targeting information to the appropriate shooter, whether the shooter is a tank division, an individual tank, an artillery battery, an individual rifle man, a naval battle group, an individual ship, an air wing/ squadron, or an aircraft in flight. This means the need to have the right shooter in the right place; locating and identifying the target correctly and quickly; allocating and assigning targets rapidly; getting the "shoot" order or general authority to the shooter; and then assessing the battle damage accurately.

At whatever the unit level, Shock and Awe are provided by the speed and effectiveness of this cycle. Then, the ability to do this simultaneously throughout the battlefield creates a strategic Shock and Awe on the opposing forces, their leadership, and populous. This simultaneity and concurrency are central tenets of imposing Shock and Awe. When the video results of these attacks are broadcast in real time worldwide on CNN, the positive impact on coalition support and negative impact on potential threat support can be decisive.

The first priority of a doctrine of Rapid Dominance should be to deter, alter, or affect the will and therefore those actions that are either unacceptable to U.S. national security interests or endanger the democratic community of states and access to free markets. These political objectives are generally those envisioned in the major and lower regional conflict scenarios (MRC & LRC). Should deterrence fail, the application of Rapid Dominance in these circumstances should create sufficient Shock and Awe to the immediate threat forces and leadership as well as provide a clear message for other potential threat partners. The doctrine of Rapid Dominance would not be limited to MRC and LRC scenarios. It has applications in a variety of areas such as countering WMD, terrorism, and perhaps other tasks. The challenge is that should deterrence fail, the execution of a response based on Rapid Dominance must be proportional to the threat, yet decisive enough to convey the right degree of Shock and Awe. Rapid Dominance cannot solve all or even most of the world's problems. We repeat our disclaimer that this is not a silver bullet. However, Rapid Dominance and its capacity for achieving Shock and Awe could be applied for egregious threats or violations of international law, such as:

- Direct military threats to the territory of the U.S., its friends, and allies;
- Blatant aggression involving a large state crushing a small state;
- Rogue leader/state sponsored terrorism/use of WMD;
- Egregious violations of human rights on a large scale; and
- Threat to essential world markets.

Clearly, the Information Highway is crossing all sovereign borders and penetrating even the most closed societies. The inequities and benefits in all societies are becoming known to the masses as well as the power brokers. The requirement for Rapid Dominance to develop sophisticated capabilities to penetrate the Information Highway and create road blocks as well as control inputs/outputs to the highway both overtly and covertly is fundamental to the concept.

These same techniques also apply to law enforcement agencies targeting international crime and drug cartels using the highway. Closer interagency cooperations and coordination between military and law enforcement activities and capabilities must be established. Experience with the military involvement in the drug war revealed considerable cultural differences between these organizations. Overcoming these cultural differences among organizations is not easy. The required trust and confidence for sharing sensitive information and support between these agencies and the military needs to be developed further. Interagency coordination and cooperation must be raised to a new level of sophistication. Some laws may need to be changed. War in Cyberspace does not recognize domestic or foreign boundaries. In this environment the subjects of Information Warfare and Information In Warfare take on new meaning and require focused development. We must become proficient within this environment.

Operational Assumptions

- The enemy picks the time and place to initiate the conflict (i.e., we are surprised).

- We then attain control of the initiative through superior speed, knowledge, and capacity to act and react.

- Our forces are perceived to be invincible; engagements must convince the enemy there is no hope.

- Combat must be unrelenting and omnipresent at times, places, and tempo of choosing.

- Allied operations must be thoroughly integrated, from political objectives through combat to include psychological warfare.

- The enemy must be hit in those areas of greatest importance to him and devastated by the ferocity and swiftness of our attack.

From these assumptions, certain operational criteria follow that help to define a Rapid Dominance Force with more specificity in improving:

- Intelligence, indications, and warning on an aggressor's actions

- The length of time required for a decision to react

- Decisive responses at various levels and times after the crises or conflict begins to develop:

    – Respond in 1 to 3 days with air and missile strikes and special forces

    – Respond in 5 to 10 days with more massive power up to and including a joint task force of corps size

    – Respond in 10 to 30 days with a second corps

The Rapid Dominance MCP

As a next step, we need to sketch out what a Rapid Dominance Force might look like for a corps-sized air, ground, sea, and space joint task force supported by necessary intelligence assets that can impose sufficient Shock and Awe to break the will of the adversary. First, this force will emphasize capabilities to maximize the core characteristics of knowledge of self, adversary, and environment; rapidity; brilliance in execution; and control of the environment.

Knowledge means more than dominant battlefield awareness. It means understanding the adversary's mind and anticipating his reactions. It means targeting those things that will produce the intended Shock and Awe. And, it means having feedback and good, timely battle assessment to enable knowledge to be used dynamically as well as to know how our forces will react.

Rapidity means moving and acting as quickly as necessary and always on a timely basis. Rapidity can be instant or as required.

Brilliance in operations means achieving the highest standards of operational competence and, through a superiority of knowledge, maintaining the ability to impose Shock and Awe through continuously surprising and psychologically and physically breaking the adversary's will to resist. This will require training and exercising of joint land, sea, air, space, and special forces to new standards of excellence and competence. It is mainly in training where the difference lies in achieving operational brilliance. This desired standard of performance can be achieved by making innovations to permit new levels of battlefield fidelity for training units and developing leaders.

Control of the environment would include complete signature control on the entire battle area out to hundreds of miles. We would control our signatures as well as what we wanted the adversary to see or hear and what we do not want the enemy to know. Destruction of the adversary's systems would begin with long-range stealthy, or "stand-off" Zero CEP weapons, extend to FOG-M type battlefield weapons to close-in systems. Small units would be able to call in "fires" for 360 degrees on a nearly instant basis.

Attacks from all aspects would be complemented by deception, disinformation, surveillance, targeting, and killing. "Pulse" weapons would be used to disarm and actively deceive the enemy through disrupting and attacking all aspects of the adversary's electronics, information, and C4I infrastructure. It is this "lay down" of total power across all areas in rapid and simultaneous actions that would impose the Shock and Awe.

The remainder, roughly a third of this Joint Task Force, would consist of traditional platforms including conventional ground, air, and amphibious forces, naval battle group forces, and the necessary supporting logistical, C4I, medical and other capabilities and ground forces to conduct and sustain conventional or traditional operations if needed and to support or defend traditionally vulnerable targets such as ports, roads, and other infrastructure.

Tactical employment is, of course, dependent on the conditions of the MRC. In general, the most rapidly deployable units of this corps, the future equivalent of the Eighteenth Airborne Corps, would be sent to secure or reinforce a limited area into which the remainder of the force would flow. This AOR would be self-protected. Our goal is that perhaps a Rapid Dominance force of as few as 2,000 troops could successfully defend against an enemy of 10-20,000 in an MRC and that a full corps can be deployed within 5 to 10 days.

These units would arrive quickly and, as directed, begin disarming, destroying, and disabling the enemy's military wherewithal using "stand-off" capabilities. Forward-based or long-range reconnaissance units could be employed/ supported by UAVs and overhead surveillance.

Units would be forward deployed in accordance with their time phased plan. These units would be used either to complete the attack or to carry it to the adversary, occupy selective territory physically, or carry out the requirements of the post-war occupation campaign. Should traditional forces be needed, they would of course be available.

Protection and self-defense would partly be provided by controlling the environment. In effect, we would cast a cloak around the adversary and permit the adversary to see and know what we alone provided. This would leave an adversary blind, deaf, and dumb. With superior and rapid firepower, the blinded, deafened enemy would be destroyed and defeated as we saw fit. This would maximize Shock and Awe and help break the adversary's will.

In OOTW, the Rapid Dominance JTF might function as follows. First, the ability to deploy dominant force rapidly to attack or threaten to attack appropriate targets could be brought to bear without involving manpower-intense or manned sensors and weapons. Second, once deployed, since self-defense is likely to be required against small arms, mines, and shoulder carried or mortar weapons, certainly some form of "armor" or protective vehicles and shelters would be necessary. However, through the UAVs, C4I, and virtual reality systems, as well as through signature management and other Shock and Awe weapons including High Powered Microwave (HPM) and "stun-like" systems, this force would have more than dominant battlefield awareness.

There are, of course, caveats. Unless strategic or policy objectives are in line with operational capabilities, military force is unlikely to be a useful instrument. It is also unlikely that any operational construct, no matter how brilliantly conceived, could overcome such a disconnect. Vietnam and Somalia remind us of these limitations.

The assimilation of intelligence—strategically, culturally, and operationally—is a central thrust and component of the knowledge aspect of Rapid Dominance. Our forces must not only fight smarter; these forces, at all or most levels, must be educated and trained differently with far more emphasis on intelligence, broadly defined. This knowledge, when applied rapidly under conditions of brilliance and in a controlled environment, is a centerpiece of Rapid Dominance.

There must be full comprehension of the adversary across strategic, political, military, cultural, intellectual, and perceptual lines. This understanding must go beyond how an adversary might use military force. Those crucial values that motivate and underlie a nation or a group must be understood if the appropriate level of Shock and Awe is to be achieved.

There are also obvious questions that must be answered. Does Rapid Dominance apply only or mostly to the high end of the conflict spectrum involving more traditional applications of force to achieve political objectives, as envisioned in the MRC and LRC scenarios? Yet to be explored is the degree to which a concept of Rapid Dominance with Shock and Awe applies to OOTW, countering terrorism against U.S. interests, controlling rogue states/leaders, etc. What are the political and military prerequisites to apply Rapid Dominance? Are they applicable and realistically achievable in the increasingly complex interaction of national non-government organizations (PVOs/NGOs) present worldwide to provide health and humanitarian care to refugees and other disenfranchised people? Would the concept of Rapid Dominance with a degree of Shock and Awe offend and generate counterproductive public relations backlash from those who believe force should only be used as a last resort and then with a measurable degree of proportionality?

At this point, we can only raise questions and expect to have them answered at a later date. This line of questions, concerns, and issues as well as a host of others, needs to be examined up front and answered in the Rapid Dominance concept development process. We must be careful that we do not overvisualize Rapid Dominance versus the reality of credible/affordable capabilities to execute the concept. Rapid Dominance must still confront the fog of war. Decisions will still be made based on judgment and confidence in the intelligence provided, the estimate of threat intentions, knowledge of true center of gravity targets, and confidence in our own force capabilities to inflict Shock and Awe. In fact, the key will be the ability to penetrate this fog with increased clarity and to control events now unmanageable through more rapid gathering, analyzing, and distributing actionable information. Complicating the issue is the fact that the U.S. has not clearly defined its role in the post-Cold War era. As the world's only credible superpower, the U.S. cannot avoid a leadership role but neither can it avoid the focused criticism applied to all leaders. This is the classical "damned if you do and damned if you don't" syndrome.

At this stage, the concept of Rapid Dominance is a work in progress. It needs to be "operationalized." By designing a nominal MCP and fitting with it paper systems and capabilities, we can explore the answers to many of the questions we raised above. Three steps are needed to proceed down the road on the way to a real capability. First, feasibility of the requisite technical capabilities needs to be established. Second, wargaming of the MCP must be done. Third, and perhaps most difficult, deriving the means for implementing the most promising aspects of Rapid Dominance must occur.

An Outline for System Innovation and Technological Integration

Achieving Shock and Awe is central to Rapid Dominance, and therefore must serve as the key organizing principle for any rigorous examination and exploitation of system concepts and technologies for Rapid Dominance. Understanding the interplay between technology and doctrine is not only or simply a straightforward matter of establishing operational requirements and then seeking to attain them through invention and design. It is a complex and interactive process of experimentation and discovery wherein intellect, hard work, endurance, and innovation must drive the use of technology. Rather than make changes, however significant, to modifying current capabilities or building newer, similar ones, Rapid Dominance seeks to identify and field systems specifically designed to achieve

Shock and Awe—systems that may break the mold much as the Model T Ford once did years ago.

The genetic decoders in bioengineering laboratories, computer-aided design tools used by engineers, vast database management systems in place in corporate offices, computer-controlled machines enabling composite materials, and the countless academic, business, and personal computers are all evidence of the prominent and ever increasing role information technologies have assumed in modern economies. Many of the technologies underlying the Information Age are being spearheaded by U.S. small business and its entrepreneurial culture. Certainly, from the huge consumer electronics firms in Japan to software development businesses in India, the rest of the world participates and competes. But few can deny that U.S. industry provides the leadership in and is the preeminent developer of information technologies as they are most broadly defined. This leadership position, properly leveraged, provides the United States with an ever increasing military advantage over competing nations.

Leveraging technology requires more than merely incorporating it into U.S. forces; it is likely to include a significant redesign of both forces and leadership to embrace these rapidly evolving technologies. Many of the technologies that will support Rapid Dominance are already discernible.

Unlike the impact of nuclear weapons, it is unlikely that a single technology or system will emerge to produce Rapid Dominance. It will only be attainable through the broadest integration of strategic concepts, doctrine, operational needs, technological advances, system design, and appropriate organization of command, control, training and education. And only a large, immensely capable country such as the U.S. may be able to achieve this.

Rapid Dominance seeks to integrate this confluence of strategy, technology, and innovation. Four core characteristics were defined earlier as crucial:

- Complete knowledge of self, adversary, and the environment;
- Rapidity;
- Brilliance of execution; and
- Control of the environment.

What follows is illustrative rather than exhaustive of how technology can be used in a broad system approach. Many of these technologies currently are being addressed within the defense community. Analysts, military strategists, acquisition planners, and even "futurists" are wrestling with the meaning and consequences of the Information Age. Our focus on systems and technologies begins with these four characteristics.

Knowledge of Self, Adversary, and Environment

In the modern threat environment, it is difficult to estimate where the next crisis may occur, let alone the next war. Even 5 years ago, who would have foreseen the significant involvement of the U.S. military in places like Somalia, Haiti, Rwanda, Bosnia, and the South China Sea? To which hot spots can we expect to see U.S. troops deployed over the next 5 years? Over the next 20? In this section we argue that, in addition to improving our force capabilities, the U.S. must develop an intelligence repository far more extensive than during the Cold War, covering virtually all the important regions and organizational structures throughout the world.

During the Cold War, intelligence agencies focused more on a bipolar world and built sizable organizations to collect information on "the other side." This same intelligence structure, in the main, is in place today facing a multipolar world, where any number of power structures—whether they be states, international organizations, or even small groups of individuals—must be monitored with an understanding that extends to their leadership, culture, economic direction, and military capability.

As the technologies relevant to knowing the adversary and his environment are examined, an emerging theme is the clear shift from technology developments that once resided within our government to those driven by commercial demands. For example, the information technologies used by U.S. intelligence agencies are of such complexity, importance, and expense that they are referred to as "national assets" and are developed and managed by large, dedicated organizations. Even here, commercial companies are rapidly encroaching on what once seemed to be an unassailable market position in Earth observation systems. One may already purchase synthetic aperture radar interferometry images from any number of sources, and panchromatic visual images with one meter resolution will soon be available over the counter for remarkably little cost. Indeed, the only real barrier to this burgeoning market is the understandable concerns that governments have with allowing such technology to be widely available. In areas such as encryption and data security, commercial developers are more likely to reach limits of government acceptance before those of technological capability.

With untold billions invested in communications systems, even the most modern U.S. military communication systems often compare poorly with commercial systems. While this has long been the case for fielded systems, it is becoming true for even the most sophisticated research and development programs being undertaken by defense organizations.

As a case in point, one may consider a program recently initiated by the Defense Advanced Research Projects Agency (DARPA) called Battlefield Awareness and Data Dissemination (BADD). At the heart of this program, large amounts of data are collected within a vast database residing on commercial computers and enterprise management systems. This information is then disseminated to the troops through the commercial Global Broadcast System (GBS) onto "set-top" boxes, an enabling technology that was developed commercially. Even with this leveraging of private industry, there is a real question as to whether DARPA will be able to field a system that would compete well with surprisingly similar commercial systems. Internet channels planned by media industry giants such as BSkyB will offer multi-megabit, interactive, digital data connections to the Net merely as an enticement for subscribers to enroll for their full digital broadcasting service (200 to 300 channels of digital video and sound). Understanding that there is much more to BADD than the little discussed here, one still almost wonders whether DARPA could simply buy a subscription and connect it to an appropriate, commercial, network management system. More to the point, if even well funded and aggressive technology development organizations such as DARPA find it difficult to remain ahead of commercial advancements, there may be a fundamental lesson to be learned regarding the management of defense-related technologies.

Knowledge and Intelligence

"Intelligence" is comprised of five categories of knowledge and understanding: a society's leadership; culture and values; the strategic, political, economic, and physical environment; military capabilities and orders of battle; and comprehensive battlefield information. Examples of technologies and system approaches of potential relevance in these areas are discussed below.

Understanding potential adversaries, coalition partners, and involved neutral countries implies an infrastructure for acquiring an in-depth knowledge about cultures, leadership values, and other driving factors that allow us, when needed and on a timely basis, to get "into their minds." Applicable technologies include automated language translators, interactive and autonomous computer simulations, advanced database systems for organizing and understanding data and transactions of individuals and institutions, and computerized educational systems for training and learning these skills.

Collecting sufficient and timely environmental information is crucial to Rapid Dominance. Logistics, demographics, and infrastructure are broad areas of collection along with geography, road/rail/ship lanes, utility sites and corridors, manufacturing, government sites, military and paramilitary facilities, population demographics, economic and financial pressure points (such as oil wells or gold mines), and major dams and bridges. Technologies used to provide environmental awareness include traditional means such as satellites that can be augmented with dynamic sensor management tools for optimizing observation routines. The vast quantities of data that reside on the world's computer networks, if properly exploited, provide another rich source of information. Data mining tools, such as Web crawlers, gatherers, brokers, and repositories that pull and organize data from public networks, will be essential to building a more complete picture of potential adversaries. Since not all databases and host computers are cooperative with these methods, offensive information warfare tools will be required to obtain specific pieces of information that are vital for national security purposes.

Once data are collected, they must be processed and disseminated and then stored for future access. Enterprise data storage and retrieval systems that are capable of working with many terrabytes (1,000 gigabytes) of information are already commonplace. Since it is impossible for humans to comprehend such vast quantities of information without some assistance, data exploitation tools (filters, fusion, automatic target recognition, image understanding, etc.) will be crucial technologies. Finally, the information, once processed, will be of little use if not disseminated to the right people in a timely fashion. "Intelligent data" dissemination and wide bandwidth communications are examples of essential technologies emerging in this area.

In addition to knowledge about regions and locations where U.S. force may be applied, it is important to maintain vigilance and up-to-date knowledge on specific "hot spots" and to have sufficient flexibility within the system to shift attention rapidly to new areas. Systems addressing this more time-sensitive set of tasks would include light, quickly deployable satellites, high altitude and endurance unmanned aerial vehicles, manned platforms, and unattended ground sensors.

As a crisis unfolds and the insertion of U.S. troops or other military action becomes more probable, information needs and the number of information consumers both increase dramatically. Information that must be collected and correlated include targeting, battle damage assessment (BDA), weather, terrain, infrastructure, tracking of special targets, logistics, position and status of our own troops, identification friend or foe (IFF), and status of material. It is vitally important that sufficient sensor systems work in all weather conditions and at night to maintain the "operations tempo" required by Rapid Dominance.

Battlefield awareness requires three information technologies: collection, fusion, and dissemination of real-time actionable information to a shooter. Rapid Dominance requires an unprecedented level of real-time information collection that will be provided by sensor systems such as space platforms, UAVs, unattended ground sensors, and advanced manned reconnaissance platforms. In addition, the entire infosphere of the adversary will be monitored not only for classic information such as operational commands but also to determine the shock effect being created by Rapid Dominance operations. Collecting data from cooperative sources such as one's own troops, allies, and friendly non-combatants is also critical. While *Operation Desert Storm* showed the value of self-location sensors such as GPS, the friendly fire casualties demonstrated that there is still work to be done in terms of giving each commander and soldier sufficient information to operate effectively. Much of this information, such as the physiological status of individual combatants, is not currently collected, and much of what is sensed is not properly disseminated.

It would be hard to overstate the importance of information dissemination within Rapid Dominance. Administering Shock and Awe requires a spectrum of attacks that the adversary is unable to fathom, but our own forces must operate effectively, even aggressively, within an environment that could easily lead to serious information bottlenecks and overload. Commercial technologies will be key to the U.S. developing a structure to effectively disseminate information. Already, commercial communications technologies such as global broadcast satellites and protocols like those underlying the Internet have been used as stop gaps by the U.S. military in major deployments.

Merely transmitting the right information at the right time will not be sufficient for operations enabling Rapid Dominance. Information will need to be fused to create knowledge-based displays. The technologies that will be important in this area go beyond the data fusion algorithms currently in place and should leverage heavily off of technologies in fields such as computer image generation, virtual reality, and advanced simulation.

Rapidity

In a technology sense, rapidity includes the speed of operational planning, determining appropriate action, deployment, and employment all focused toward minimizing response time. Three factors combine to make military planning far more difficult today than in the Cold War era. First, there is great uncertainty early on in the location of a conflict, who the adversary may be, and with whom one may be allied. Second, there is normally very little time available for planning, with the military sometimes having only weeks or days before committing troops to an unanticipated mission. Third, vastly more information is available to the planner, which is both a blessing and a curse. Several technologies that partially define Intelligent Dynamic Planning will make it easier for the commander to plan Rapid Dominance:

- Model based planning
- Machine intelligence
- Dynamic planning based upon feedback and new information
- Selectively automated decision aides (commanders associate)
- Imbedded rehearsal and training

Brilliance in Execution

It is impossible to institutionalize brilliance. However, the standard can be set. The Dynamic Planning noted above is part of the capability for this characteristic as are the systems and technologies discussed below.

Technologies Critical to Achieving Brilliance in Rapid Dominance

For shock to be administered with minimum collateral damage, key targets of value must be neutralized or destroyed, and the enemy must be made to feel completely helpless and unable to consider a meaningful response. Furthermore, the enemy's confusion must be complete, adding to a general impression of impotence. Most importantly, strategic targets, military forces, leadership and key societal resources must be located, tracked, and targeted. This will require substantial sensor, computational, and communication technologies. Designated targets must be destroyed rapidly and with assurance. Finally, the status and position of friendly forces must be known at all times, and the logistics supporting them must be sufficiently flexible to allow for rapid movement, reconfiguration, and decentralization of location.

Several technologies that can help in this are discussed below, as divided into the following subsections: sensors, computational systems, communications and system integration.

Sensor Technologies

Sensor technologies are grouped into four areas: active, passive, imbedded, and processing.

Active sensors: By far, the most important of the energy-emitting sensors is radar. Among the best all-weather capabilities of any type of sensor, the role for and capabilities of radar have steadily increased since the Second World War. Radar systems are used for early warning, air defense, air asset management, air traffic control, naval fleet defense, detection and tracking of moving ground targets, missile targeting, missile terminal guidance, terrain data development, and weather prediction. For Rapid Dominance, radars and other active sensors must operate with low probability of intercept. Particularly with stealthy systems, this will present a unique challenge to military systems where one may not expect a great amount of "spin-on" from the commercial sector. It is vitally important to be able to sense the enemy under all conditions and environments. Sensors must penetrate foliage and walls and detect threats such as underground and underwater mines.

There are many other important active sensor classes, three of which are active acoustics, lidar and magnetic anomaly detectors. Broadband underwater active acoustics could address pressing needs such as shallow-water anti-submarine warfare and mine detection (both buried and silt covered). The practical application of lidar is a relatively recent development enabled by advances in laser, power management, and data processing technologies. Lidar can be used for fire control, weapon guidance, foliage penetration (vegetation is translucent in the near infrared (NIR) regime), and target imaging/ recognition. Lidar detects shape directly and shape fluctuations such as vibration and motion and has proven very hard to spoof. Magnetic anomaly detectors will continue to find application in areas of anti-mine and anti-submarine warfare and in screening for weapons at security checkpoints and elsewhere.

Electronic emissions are of themselves a liability only where they create a signature of use to an enemy. The ability to emit energy, yet in ways that are less discernible, should be an attractive avenue to explore for the future. The coordinated application of many sensor platforms, some of which may be completely passive, in conjunction with emitting sensors is a potentially major area of exploration.

Passive sensors: Among the passive sensor types, the most important for U.S. forces is forward-looking infrared (FLIR). FLIR technology has allowed the U.S. to "own the night," as was handily displayed in *Operation Desert Storm*. Some of the significant technology advancements underway in this area include multiple wavelength sensors, very large focal planes, and the increasing performance of uncooled sensors. Particularly in the area of uncooled sensors, commercial developments are underway that promise to drastically reduce the cost of competent IR sensors.

Other passive sensor technologies of note include hyperspectral visible/NIR collection and processing and inexpensive, scatterable, unattended ground sensors (acoustic, seismic, "hot spot," etc.). Hyperspectral imaging allows target searches to be conducted in the frequency domain, as opposed to the spatial domain as is the norm today. This provides a powerful new input for automatic target recognition (ATR) systems, is useful for addressing low observables (LO), and is especially important for remote imaging assets.

Unattended ground sensors allow critical areas to be monitored continually. For example, the actual area of operations for Scuds in ODS was relatively small, but it was very difficult for then-current sensing systems to oversee. Technologies being developed in the area of microelectromechanical systems, in particular, hold promise for enabling capable and inexpensive sensor fields.

Imbedded sensors: Monitoring the position and status of Blue and friendly forces and assets is of equal importance in tracking the enemy. GPS presented a tremendous advantage to troops in ODS. This capability needs to be extended down to the individual soldier, and the status of all critical material and personnel needs to be tracked.

Sensor signal processing: Finally, the signals from modern sensors are of limited use without proper processing and presentation to the user. This area will be developed further in the computational technologies section. Technologies that are historically grouped with sensor systems include automatic target recognition, imbedded multisensor fusion and correlation, and displays.

Computational Technologies

The capabilities of the integrated circuit (IC), and in particular the microprocessor, continue to increase unabated. Certainly, physical limits must be approached at some point, but each looming barrier has so far been met by technological innovation. Nevertheless, should the march of IC improvements slow somewhat, the software and networking technologies that are being developed at an accelerating pace will permit the vision of Rapid Dominance to become of ever increasing utility.

Rapid Dominance requires the collection, management, and fast access of enormous quantities of information. Technologies that will enable this include computational hardware advances such as increasingly powerful workstations, reduced-cost image generators, massively parallel machines, compact displays, reduced-cost memory devices (i.e., DRAM, RAID, and optical jukeboxes) client/server-specific database engines, reconfigurable simulation cells, "wearable" PCs, advanced human-computer interface (HCI) techniques (i.e., voice interfaces and those coming to define "virtual reality"), and PCMCIA technology for peripherals (i.e., digital comms boards, miniaturized hard drives, and modems).

Software advances will be even more critical for Rapid Dominance. Areas of importance include:

- Network data engines
- Object-oriented architectures
- Advanced modeling and simulation

- Machine intelligence
- Automatic target recognition
- Computer-aided software engineering (CASE) tools

Network technologies are just now emerging but are being driven at a frenzied pace in the commercial marketplace. A variety of advanced tools beyond "hot link" browsing are being introduced daily. Data browsers, brokers, gatherers, and network repositories are being released, as demonstrated by products like *Harvester* and *Netscape's Catalog Server*. Platform independent languages such as JAVA and their associated virtual computational engines promise the same network flexibility for programs that is now enjoyed by data.

Perhaps the most important area of technology development for Rapid Dominance is the development of practical object-oriented architectures and protocols. Protocols such as CORBA, OLE, ALSP, HLA and DIS[1] are changing

---

[1] CORBA (common object request broker architecture), OLE (object linking and embedding), ALSP (aggregate level simulation protocol), HLA (high-level architecture), DIS (Distributed Interactive Simulation). These are all protocols or the architectures defining protocols that, in part, enable disparate software and/or hardware components to be linked or otherwise share information and logical elements.

the face of computing, making it much easier to link programs and databases, and access and correlate information that was previously "entombed" within its legacy application.

One interesting application area migrating toward an object-oriented approach is geospatial databases. In the past, geospatial data were stored as either raster-based or vector information, and significant processing was required for users to make queries regarding roads, areas, or objects such as building sites. A new approach, called a spatial database engine, creates intuitive objects from standard geospatial databases and uses commercial databases to add attributes to the objects. This is a very powerful technique that allows geospatial data, a key element of warfighting, to be managed quickly and efficiently using commercial-off-the-shelf (COTS) software. It is particularly useful for distributed databases such as one would find on a network.

Modeling and simulation is also benefiting from object-oriented technologies. Simulations were once stand-alone codes. If one wanted to simulate a joint battle, one began with an existing model (i.e., land combat) and then modified it to include other components (i.e., aircraft and ships). Similarly, if a new technology were to be modeled, new code normally had to be written, even in cases where good, validated, stand-alone technology models existed. The obvious drawbacks to this approach are that it is costly, often produces inferior simulations for the new additions, and quickly results in extremely large codes with commensurate large code management problems. Object-oriented approaches allow models and simulations to be linked to form a richer environment for examining new technologies and joint force structures.

Linking force-on-force simulations with design tools such as computer-aided design (CAD) programs and physics-based simulations presents a new type of tool referred to as simulation-based design. Once fully realized, this capability will allow new technologies to be much more easily evaluated, introducing a source for greater efficiency into today's somewhat haphazard acquisition system.

Simulations based on object-oriented architectures also promise more flexibility that will enable scenarios and unexpected situations to be made as inputs and simulated rapidly, forming the core for a battlefield visualization system capable of modeling "what if" situations. Outputs from these simulations could be used for mission rehearsal. Even today, pilots and special operations forces often "fly through" crude, three-dimensional renderings of a mission area to familiarize themselves with information such as surface-to-air missile (SAM) sites and landmarks.

The promise of computational technologies brings with it potential vulnerabilities that must be protected against threats. In a world where information plays a vital role in warfare, information collection and processing tools will become targets. Defenses against information warfare must be developed. The threat is real and is growing especially in the commercial and private sectors. Even today, malicious hackers devise data-destroying viruses and distribute them through a plethora of electronic media; numerous sites on the Net are dedicated to the discussion and development of offensive computer viruses, with ample tools for even the novice to download and employ. Moreover, computer crimes cost the world economy billions of dollars annually. Although information warfare poses serious threats, the realm of information is where operations underlying Rapid Dominance most reside, and the enemy will find himself fully engaged should he choose to fight on our terms. Rapid Dominance is essentially information warfare on a grand scale in all dimensions of offensive, defensive and leveraging effective use of available information.

Communication Technologies

One of the modern communication devices being fielded within U.S. forces today is the SINGCARS radio. With a data rate of somewhat less than 10 kbps, SINGCARS is woefully inadequate for supporting Rapid Dominance. However, more appropriate technologies are emerging:

- GBS and other satellite broadcast services
- Wider bandwidth, digital communication protocols
- Asynchronous transfer mode (ATM) switches
- Advanced comm relay platforms (UAV, Lightsat, Iridium, etc.)

GBS, for example, figures prominently in the BADD (battlefield awareness and data dissemination) program that aims at providing close to 30 Mbps of data broadcast bandwidth. This will be supported by multi-terrabyte databases, advanced data browsers, and query managers, and will be linked to the Joint Tactical Internet.

Networking must also be supported by communications technologies. The basic problem of a battlefield network is that while some nodes may support very large data pipes, a number of nodes will be operating at SINGCARS data rates. This led to the BADD notion of one-way data broadcasting via GBS of large data files (such as

UAV video and overhead imagery) and very low bandwidth data querying back to the data sources.

Modern communications will tend to be more multimedia-based, which is particularly important for Rapid Dominance, where decisions must be made quickly based upon very large quantities of data, some of which will be collected and transmitted in real time. Technologies such as digital video teleconferencing, virtual whiteboards, and even 3D virtual environments where commanders may participate in collaborative planning sessions will become important.

Finally, battlefield communications must be secure and, where feasible, non-observable to the enemy.

Control of the Environment

The actual attack of targets in order to induce Shock and Awe may, in some sense, be considered a subset of controlling the enemy's perception. It will not always be necessary to destroy numerous targets in order to induce shock. However, it would be vitally important to give the appearance that there are no safe havens from attack, and that any target may be attacked at any time with impunity and force. Furthermore, as discussed earlier, confusion must be imposed on the adversary by supplying only information which will shape the adversary's perceptions and help break his will. Finally, the enemy must be displaced from selected key positions, for if he is allowed to occupy those areas that he considers strategically important, it is difficult to imagine how his shock could be complete.

Controlling an enemy's perception of the battlespace includes manipulating his view of the threat, his own troops and status, and the environment in which he operates. This will be accomplished by selectively denying knowledge to the enemy while presenting him with information that is either misleading or serves our purposes. Sensing and feedback of an enemy leadership's perception of the situation will be critical.

Technologies of interest here include those that allow systems and entire force units to modify their signature from being very stealthy to being completely obvious. An ability to attack enemy information systems will also be critical, encompassing system technologies from laser-based counter sensor weapons to embedded computer viruses, commonly referred to as Trojan Horses. In all cases, the goal will be to deny the enemy any information that would be useful to him and to impose a construct of deception and misinformation at all levels of operations.

Clearly, technologies necessary to achieve battlefield awareness already mentioned will be crucial in allowing a "perception attack" (a form of information warfare) to be successfully carried out. The need and requirements for Battlefield Damage Assessment (BDA) will increase dramatically. It will be necessary to understand not only whether a target was killed but also how enemy leadership, troops, and society viewed this destruction.

So far, primarily information technologies have been discussed. Obviously, there will continue to be requirements for numerous other types of systems. Among the more important system technologies critical to achieving control of the environment include:

- Weapons platforms with stealth technology
- Weapons systems
- Robotic systems

Weapons platforms

One of the fundamental rationales for weapons platforms is to move people and ordinance to within an effective range of the target. Centuries before smart weapons and robotic systems, this reasoning was understood intuitively. Since ordinance must still be placed on the target, weapons platforms such as described below still demand consideration.

- Stealthy bombers and strike aircraft either land or sea platform based
- Arsenal ships
- Submarines with conventional cruise missiles
- Stealthy land vehicles
- Stealthy observation/attack helicopters

Stealth, combined with stand off, will contribute strongly to the protection of manned systems on the modern battlefield and will also be used extensively for other, high-value unmanned systems. However, protection of the force is inherent within the concept of Rapid Dominance, and it will rely upon the control of information and the enemy's perception of events, stealth being one of the elements enabling this control.

Weapons systems

Smart munitions will be required on the future battlefield. Linked with information technologies, the combination will allow killing any target that can be identified. The main element Rapid Dominance requires of weapons systems is the ability to be rapidly focused on objectives as identified and targeted by commanders using the information management systems already discussed. Commanders will require the flexibility to call massive, precision strikes or to attack individual, high-priority targets with near zero CEP. This implies a mixture of weapons comprised of systems such as those mentioned below.

- Cruise missiles
- Zero CEP, long-range cruise missile ("President's weapon")
- Stand-off submunition platforms
- Smart submunitions
- Brilliant submunitions
- Wide area smart mines
- Long-range and short-range surface attack missiles

Robotic systems

Robotic systems are an important area of consideration within Rapid Dominance. First, selected robotic systems will enable the force by making it more responsive in concentrating sensors and weapons. Second, they will make fighting a 24-hour battle feasible even with reduced manpower within the force structure. Third, robotic systems can provide force presence even in areas considered too dangerous for a large manned element. Finally, since the ultimate operational goal of Rapid Dominance is to create shock, one may consider the effect that fighting robotic systems may have on the enemy.

In examining the utility of robotic systems within Rapid Dominance, one must first consider that, by any measure, robotic systems have not lived up to the optimistic expectations placed on them in the past. From the overburdening of the Aquilla UAV to the massive and poorly planned investment in robotics made by General Motors in the early 1980s, robotics has been an area of unfulfilled promises. However, the reasons for a string of spectacular failures lie more with planners' faulty attempts to understand and incorporate the technology than by egregious shortcomings of the technology itself. Robots have been seen as replacements for manned systems rather than extremely complicated and capable machines suitable for a set of tightly defined tasks. Robotic systems, or taskable machines as some are beginning to refer to them, hold promise for the future simply because they represent the intersection of a myriad of fast-moving technology areas such as information technologies, communications, microelectronics, micro-electromechanical systems, simulation, and computer-aided design and manufacturing. In some sense, taskable machines are the physical embodiment of information technologies. It may well be that in the future the joke will be, "Never send a robot to do a man's job." But even so, there will be ample jobs for taskable machines and the society that learns to properly design, build, control, and integrate these systems into their force structure will gain significant advantage over any potential opponent.

Conclusion

The technologies and systems presented in this section are not extraordinary nor do they comprise a complete list. Indeed, entire fields such as materials, bioengineering, and microelectronics are left for future consideration, although they are of obvious and vital importance. Also not addressed here are the training, education, and organizational implications required under a regime of Rapid Dominance. Given the overriding importance of information collection and management, these will need to be addressed across the defense community as it is most broadly defined.

Rapid Dominance combines a doctrine and operational concept that challenges the current process of how new technologies invented in the commercial sector are incorporated into defense, and provides an affirmative methodology for research, development, and system integration. We must learn to exploit the potential of these technologies even though, in many cases, this development process in the private sector is profoundly independent from how we conduct the business of defense. It is this environment of innovative upheaval that any useful foundation for strategic and operational thought must address. Rapid Dominance capitalizes on, and may even require, this rapid and chaotic development of technology.

We believe that what will distinguish Rapid Dominance from other doctrines is first that it uses an intellectual construct to drive innovation and innovation to drive exploiting and integrating technology into new and perhaps somewhat differently constructed systems. Second, it is the comprehensive quality of Rapid Dominance in which strategies, doctrine, technology, systems, operations, training, organization, and education are dealt with together that may make the most significant difference. But, as the reader will discern, specific identification and design of Rapid Dominance systems is part of the next step.

Future Directions

At this stage, Rapid Dominance is an intellectual construct based on these key points. First, Rapid Dominance has evolved from the collective professional, policy, and operational experience of the study group covering the last four decades. This experience ran from Vietnam to *Desert Storm* and from serving with operational units in the field to being part of the decision-making process in the Oval Office in Washington. It also included immersion in technology and systems from thermonuclear weapons to advanced weapons software.

Second, Rapid Dominance seeks to exploit the unique juncture of strategy, technology, and innovation created by the end of the Cold War and to establish an alternative foundation for military doctrine and force structure.

Third, Rapid Dominance draws on the strategic uses of force as envisaged by Sun Tzu and Clausewitz to overpower or affect the will, perception, and understanding of the adversary for strategic aims and military objectives. But, in Rapid Dominance, the principal mechanism for affecting the adversary's will is through the imposition of a regime of Shock and Awe sufficient to achieve the aims of policy. It is this relationship with and reliance on Shock and Awe that differentiates Rapid Dominance from attrition, maneuver, and other military doctrines including overwhelming force.

Shock and Awe impact on psychological, perceptual, and physical levels. At one level, destroying an adversary's military force leaving the enemy impotent and vulnerable may provide the necessary Shock and Awe. At another level, the certainty of this outcome may cause an adversary to accept our terms well short of conflict. In the great middle ground, the appropriate balance of Shock and Awe must cause the perception and anticipation of certain defeat and the threat and fear of action that may shut down all or part of the adversary's society or render his ability to fight useless short of complete physical destruction.

Finally, in order to impose enough Shock and Awe to affect an adversary's will, four core characteristics of a Rapid Dominance-configured force were defined. First, complete knowledge and understanding of self, of the adversary, and of the environment are essential. This knowledge and understanding exceed the expectations of dominant battlefield awareness and DBA becomes a subset of Rapid Dominance.

Rather like the wise investor and not the speculator who is only familiar with a particular company and not the stock market in general, the Rapid Dominance force must have complete knowledge and understanding of many likely adversaries and regions. This requirement for knowledge and understanding will place a huge, new burden on the military forces and necessitate fundamental changes in policy, organization, training, education, structure, and equipage.

Second is rapidity. Rapidity combines speed, timeliness, and agility and the ability to sustain control after the initial shock. Rapidity enables us to act as quickly as needed and always more quickly than the adversary can react or take counter-actions. Rapidity is also an antidote to surprise. If we cannot anticipate surprise, or are surprised, rapidity provides a correcting capacity to neutralize the effects of that surprise.

Third, and most provocatively, is setting the standard of operations and execution in terms of brilliance. The consequences and implications of setting brilliance as the standard and achieving it are profound. Reconfiguration of command authority and organization possibly to decentralization down to individual troops must follow. Allowing and encouraging an operational doctrine of the "first to respond" will set the tempo provided that effective de-confliction of friendly on friendly engagements has been assured.

This, of course, means that complete revision of doctrine, training, and organization will be required. The matter is not just "fighting smarter." It is learning to fight at even higher standards of skill and competence.

Fourth is control of the environment. Control is defined in the broadest sense: physical control of the land, air, sea, and space and control of the "ether" in which information is passed and received. This requires signature management throughout the full conflict spectrum—deception, disinformation, verification, information control, and target management—all with rapidity in both physical and psychological impact. By depriving an adversary of the physical use of time, space, and the ether, we play on the adversary's will and offer the prospect of certain destruction should resistance follow.

The next step in this process must be specifically defining this Rapid Dominance force in terms of force structure, capabilities, doctrine, organization, and order of battle. We have begun this effort and are focusing on a joint task force sized somewhere between a reinforced division and a full corps (i.e., a strength of 75,000 - 200,000). We also have the aim of being able to deploy this force within 5 to

10 days of the order to move and, of course, will be able to send smaller force packages on a nearly instantaneous basis. We appreciate the mobility and logistical implications of this requirement.

Once we design this "paper" force and equip it with "paper" systems, we must evaluate it against the five basic questions and tests we noted in the Prologue.

The first test of this Rapid Dominance force will be against the MRC. The comparison, in the broadest sense, must be with the programmed force and whatever emerges from the Quadrennial Defense Review of 1997. We will need to examine closely how and where and why Rapid Dominance and Shock and Awe work and where they do not. At the very least, we expect that this will help strengthen the current force and improve current capabilities. Of course, it is our hope that this test will validate Rapid Dominance as a legitimate doctrine.

Second, the Rapid Dominance force must be tested across the entire spectrum of OOTW. These are the most difficult tests because, in some of them, no force may be suitable and no force may work.

Third, the test of determining the political consequences of Rapid Dominance must be conducted. On one hand, if this force capability can be achieved and Shock and Awe administered to affect an adversary's will, can a form of political deterrence be created? In the most approximate sense, and we emphasize approximate, the analogy with nuclear deterrence might be drawn. An adversary may be persuaded or deterred from taking action in the first instance. On the other hand, this capacity may be seen as politically unusable and allies and others within the United States may not be fully trusting of the possessor always to employ this force responsibly.

Fourth is the test of the implications of Rapid Dominance for alliances and for waging coalition warfare. Our allies are already concerned that the United States is leaving them far behind in military technology and capability. If we possess this force and our allies or partners do not, how do we fight together? Our view is that this can be worked out through technology sharing and perhaps new divisions of labor and mission specialization. However, these are important points to be considered.

Finally, what does all this mean for resource investments in defense?

It is also likely that because Rapid Dominance will cause profound consequences, the iron grip of the political bureaucracy will make a fair examination difficult. It is no accident that other attempts at change, especially those that ask for or are tainted with reform, have had a short life span. It is interesting to note in this regard that the President's Commission on Intelligence and its fine report that recommended changes and refinements to the U.S. intelligence community, despite a very positive initial reception, led to only a few meaningful actions.

This discussion leads to two final points. We are all too well aware that any strategy and force structure have vulnerabilities and potential weaknesses. The experiences that this study group collectively had in Vietnam makes this concern very strongly held. We observe that in the private sector, the vulnerability of information systems is real and is being exploited. A former director of the FBI has told us that in New York, for example, the number one recruiting target for organized crime is the teenage computer whiz. We think that this "hacking," writ large in the private sector, must be assumed as part of the defense problem. Hence, sensitivity to vulnerabilities must be even greater, perhaps ironically, than it was during the Cold War, because exploitation can come from many more sources in the future.

Second, wags may criticize Rapid Dominance as attempting to create a "Mission Impossible Force."

To be sure, we emphasize and demand brilliance as the operational goal. However, we also know that the military today is seen as a leading example of the best American society has to offer. We wish to build on this reality. We note the experience and the performance, albeit under highly unusual circumstances, of *Desert Storm*. We see no reason why that level of performance cannot be made a permanent part of the fabric of the American military.

Because we have entered a period of transition in which we enjoy a dominant military position and a greatly reduced window of vulnerability, this is the right time for experimentation and demonstration. Rapid Dominance is still a concept and a work in progress, not a final road map or blueprint. But the concept does warrant, in our view, a commitment to explore and an opportunity that could lead to dramatically better capabilities.

We believe that through Rapid Dominance and the commitment to examine the entire range of defense across all components and aspects, a revolution is possible. If Rapid Dominance can be harnessed in an affordable and efficient way and an operational capability fielded to impose sufficient Shock and Awe to affect an adversary's will, then this will be the real Revolution in Military Affairs. We ask those who are intrigued by this prospect to join us.

Thoughts on Rapid Dominance

Admiral Bud Edney, USN (Ret.)

Why the need for a concept of Rapid Dominance? The answer lies in the combined realities of modern technology, economics, and politics.

Technology

The evolution or revolution of information technology is impacting everything we do and how we do it on a worldwide basis. The far-reaching effects of the resulting information highway that crosses all boundaries are already impacting the strategic decisions, economics, and politics of the world of nation states. Borders are no defense for the penetration of information even in highly controlled or authoritarian societies. Similarly, the exploration and use of high technology in space, together with the advent of sophisticated highly accurate ballistic and cruise missiles, means borders between states are not as important for strategic and impenetrable defenses in depth as they used to be. The rapid advancements in telecommunications technology, combined with the exploration and use of space vehicles to saturate a world hungry for information, means that leaders can no longer shield their people from the outside world. Thus information will penetrate whatever curtain or wall that is erected in a futile attempt to block it out. New centers of gravity are being created as are new vulnerability choke points. The country or power structure that harnesses the capabilities and dimensions of the information revolution as it applies to issues of national security will remain in control of its own destiny. The United States possesses a qualitative and quantitative lead that, when combined with a properly focused and coordinated (harmonized) industry, defense, and national security policy, should ensure success for the foreseeable future. Harnessing information technology and applying it to new strategic and doctrinal thought in application of military force is the essence of Rapid Dominance.

Economics

With the end of the Cold War and the dismantling of the Soviet Union, there is no major power capable of destroying the U.S. mainland. Given this absence of devastating threat, defense expenditures will continue to be squeezed to address more pressing domestic priorities. Voter demands for a balanced budget, national health care, social security reform, educational reform, family values, crime and drug use reduction, lower taxes, etc., will combine to put increasing pressure on the defense bottom line in the out years. The result will be a steady decline in war fighting readiness and force structure that will place our security interests at risk unless we leverage our technology leadership to achieve military advantage with lower force levels but increased war fighting effectiveness. This is also the essence of Rapid Dominance.

Politics

The reality of current politics is that the trauma of Vietnam, the results of the Gulf War, and our status as the only remaining superpower after the Cold War equate to some new constraints (real or perceived) on the application of military force to support our foreign policy. These political sensitivities need to be understood up front and include the following:

- The U.S. is not the world's policeman

- Involvement of U.S. Forces must be justified as essential to vital U.S. security interests

- Support of Congress and People is a necessary prerequisite
- Avoid commitment of ground forces
- Offer instead U.S. intelligence, air lift, sea lift, logistics support, etc.
- Avoid risk of loss of U.S. lives at almost all costs
- Ensure decisive force applied for mission assigned
- Rules of Engagement allow U.S. forces to defend themselves aggressively
- Minimize civilian casualties, loss of life, and collateral damage
- Specify achievable mission objectives up front with an end in the not-too-distant future sight before committing
- U.S. led coalition force preferred—U.S. Forces remain under U.S. Command. These political restraints may limit the application of Rapid Dominance to Major and Minor Regional Conflicts. This is an issue that needs further exploration and analysis.

What is Rapid Dominance?

Rapid Dominance is the full use of capabilities within a system of systems that can decisively impact events requiring the application of military/defense resources through affecting the adversary's will. Rapid Dominance envisions execution in real or near real time to counter actions or intentions deemed detrimental to U.S. interests. On one end of the spectrum, Rapid Dominance would introduce a regime of Shock and Awe in areas of high value to the threatening individual, group, or state. In many cases the prior knowledge of credible U.S. Rapid Dominance capabilities would act as a deterrent. Rapid Dominance would ensure favorable early resolution of issues at minimal loss of lives and collateral damage. The concept ideally should be able to impact adversarial situations that apply across the board, addressing high-, mid-, low-, and no-technology threats. Some of these aims may not be achievable given the political and technology constraints, but need to be explored.

Rapid Dominance expands the art of joint combined arms war fighting capabilities to a new level. Rapid Dominance requires a sophisticated, interconnected, and interoperable grid of netted intelligence, surveillance, reconnaissance, communications systems, and data analysis to deliver in real time, actionable information to the shooter. This network must provide total situational awareness and nodal analysis that enables U.S. forces to act inside the adversary's decision loop in a manner that on the high end produces Shock and Awe among the threat parties. Properly detailed nodal analysis of this grid of knowledge and vulnerability will enable the shutting down of specific or all essential functions nearly simultaneously. We expect that through these netted pieces of data, often, the sum of the parts will yield profound battlefield advantages to the possessor. The "Rapid" part of the equation becomes the ability to get real time actionable targeting information to the shooter, whether the shooter is a tank division, an individual tank, an artillery battery, an individual rifleman, a naval battle group, an individual ship, an air wing/squadron, or an aircraft in flight. At whatever unit level, Shock and Awe are magnified by the speed and effectiveness of targeting. The ability to achieve Rapid Dominance simultaneously throughout the battlefield will create strategic Shock and Awe on the opposing forces, their leadership, and society. When the video results of these attacks are broadcast real time worldwide on CNN, the positive impact on coalition support and negative impact on potential threat support can be decisive.

The top priority of Rapid Dominance should be to deter, alter, or affect those actions that are either unacceptable to U.S. national security interests or endanger the democratic community of states and access to free markets. These political objectives are generally those envisioned in the major and lesser regional conflict scenarios (MRC & LRC). Should deterrence fail, the application of Rapid Dominance should create sufficient Shock and Awe to intimidate the enemy forces and leadership as well as provide a clear message for other potential aggressors. Rapid Dominance would not be limited to MRC and LRC scenarios. It has application in a variety of areas, including countering WMD, terrorism, and other political problems. The challenge is that should deterrence fail, the execution of a response based on Rapid Dominance must be proportional to the threat yet decisive enough to convey the appropriate degree of Shock and Awe. Rapid Dominance cannot solve all or even most of the world's problems. It initially appears that Rapid Dominance should be applied sparingly for egregious threats or violations of international law, such as:

- Blatant aggression involving a large state crushing a small state

- Rogue leader/state sponsored terrorism/use of WMD

- Egregious violations of human rights on a large scale

- Threat to essential world markets

Clearly the Information Highway is crossing all sovereign borders and penetrating even the most closed societies. The inequities and benefits in closed societies are becoming known to both the public as well as the bosses. The requirement for Rapid Dominance to develop sophisticated capabilities to penetrate the Information Highway and create road blocks as well as control input/outputs to the highway both overtly and covertly is fundamental to the concept.

These same techniques also apply to law enforcement agencies targeting international crime and drug cartels using the highway. Closer interagency cooperation and coordination between military and law enforcement activities and capabilities must be established. Experience with the military involvement in the drug war revealed considerable cultural differences between these organizations. Overcoming these cultural differences is not easy. The required trust and confidence for sharing sensitive information and support between these agencies and the military needs to be developed further. Interagency coordination and cooperation must be raised to a new level of sophistication. Some laws may need to be changed. War in Cyberspace does not recognize domestic versus foreign boundaries. In this environment the subjects of Information

Warfare and Information In Warfare take on new meaning and require focused development. We must become proficient within this environment.

This breakdown of traditional boundaries requires a great deal more thought with regard to the issues of security, vulnerabilities (their's and our's), and the concept of Rapid Dominance. Does Rapid Dominance apply only or mostly to the high end of the spectrum, involving more traditional applications of force to achieve political objectives as envisioned in the MRC and LRC scenarios? Yet to be explored is the degree to which a concept of Rapid Dominance applies to OOTW, countering terrorism against U.S. interests, controlling rogue states/leaders, etc. What are the political and military prerequisites to apply Rapid Dominance? Are they applicable and realistically achievable in the increasingly complex interaction of national governments/law enforcement organizations and international as well as local private venture or non-government organizations (PVOs/NGOs) present worldwide to provide health and humanitarian care to refugees and other disenfranchised people? Would the concept of Rapid Dominance offend and generate a counterproductive public relations backlash from those who believe force should only be used as a last resort and then with a measurable degree of proportionality?

At this point, one can only raise these types of issues to be addressed at a later date. This line of questions, concerns, and issues, as well as a host of others, needs to be raised up front during the concept development phase of the development of specific Mission Capability Package concepts. We must be careful that we do not overvisualize Rapid Dominance versus the reality of credible/affordable capabilities to execute the concept. Rapid Dominance does not eliminate the fog of war. Decisions will still be made on the leader's judgment and confidence in the intelligence provided, the estimate of threat intentions, knowledge of true center of gravity targets, and confidence in our own force capabilities to inflict Shock and Awe. In fact, the ability to penetrate this fog is the key to Rapid Dominance. Complicating the issue is the fact that the U.S. has not clearly defined its role in the post-Cold War era. As the world's only credible superpower, the U.S. can not avoid a leadership role, but neither can it avoid the focused criticism applied to all leaders. We are in the classical "damned if we do and damned if we don't" syndrome. One of the serious side effects of Rapid Dominance could be that if you adapt a strategy of Rapid Dominance and succeed, you may now own the problem and be responsible for the solution. Do we know the funding tail to such a policy and are we as a nation ready to accept this cost when/if Rapid Dominance is applied in situations that are less than of vital interest? This subject needs further development beyond the limitations of this book.

Rapid Dominance and The Future Battlefield

What will the battlefield of the future really look like? The *Desert Storm* conflict indicated to many who analyzed it that the real focus of battle will no longer be force on force as we have traditionally considered it. By the time the Allied Forces engaged the opposing Iraq forces, the enemy force for all practical purposes had already been demoralized and smashed. This was accomplished by establishing air superiority followed by a carefully orchestrated campaign of precision air strikes (including Tomahawk missiles). The Iraqi ground forces were isolated by cutting off logistic support, severing communications with its leadership, and stinging them with the Shock and Awe achieved by B-52 strikes on the entrenched Iraqi forces in the open desert. Shock and awe were introduced in the manner that stealth aircraft penetrated enemy air defenses and surgically attacked center of gravity targets with impunity. Shock and awe were also present in the degree that coalition forces owned the night and could rapidly maneuver large units in terrain thought to be foreign, imposing, and unforgiving for the predominantly U.S. forces. Instead, as Colin Powell noted, the Coalition Forces cut off the head and life lines to the Iraqi Army in the field and then set about killing it. The fact that a democratically led coalition could choose not to massacre the remnants of Iraq's army during its panic-induced retreat underscores that we knew how much power we had and could employ restraint. The impact of real-time video media coverage of these events, beamed simultaneously into government headquarters and civilian living rooms worldwide, is a phenomenon that impacted events on the battlefield and further highlighted the compassion of that decision. In dealing with a "butcher" we could not fall to that level.

The battlefield of the future will not be a neat 200x200 mile box where you will know everything that is going on inside the box (although that would be an extremely helpful first step). The battlefield of the future will encompass every pressure point that controls or influences the elements of the battle. In examining this battlefield and the application of force and Shock and Awe, we seek to mass devastatingly accurate and simultaneous firepower on critical nodes/targets that count for the mission at hand, rather than necessarily having to mass large armies in the field to engage one another. Clearly, the Gulf War raised warfare to a new level with the demonstrated effectiveness and application of air to ground/water and surface to ground/water launched precision guided weapons. No longer will commanders count sorties and tonnage of ordnance dropped, but rather targets destroyed per sortie! Note: there may well be an issue of affordability here. We may not be able to get 1) high tech, 2) MRC/OOTW, and 3) large armies. This does not eliminate the requirement for sufficient force in the field to defend against an all-out assault or eject another force and occupy the contested land to ensure the objectives of conflict are carried out. Air power can punish, simultaneously destroy center of gravity targets, and so demoralize the opposing forces that land campaign objectives can be achieved with smaller forces. In some cases, the Shock and Awe achieved by the air campaign may result in an early cessation of conflict before the land campaign is necessary. This is more likely against a modernized, developed state than an underdeveloped government.

The confluence of several technologies, including all aspects of stealth aircraft, satellite global positioning, improved weapon targeting and terminal guidance, cruise missile technology, space relayed command & control, real-time surveillance from space, the introduction of JSTARS, and massive application of night vision techniques, are the first phase of these changes. With elements of this technology now more and more on the open market to whomever has the cash or friends, the advantage of obtaining greater situational awareness and real-time processing of available data cannot be taken for granted.

In future environments, and short of all-out war, it is clear that political and military decision making will have to establish close control of the actionable information distributed to shooters in the field. It is legitimate to ask why Israeli forces that had air superiority, UAV surveillance, and extremely accurate firepower capabilities in the most recent incursion into Southern Lebanon against Hezbolla terrorist attacks had to respond with an artillery barrage to one Kaytusha rocket fired from close to a known UN encampment. When this artillery response resulted in killing more than 100 refugees fleeing the Israeli operation, the result was a public relations disaster and mission failure for the stated limited Israeli objectives. This represents a case of ill-conceived application of Rapid Dominance that resulted in counter-productive Shock and Awe generating adverse public opinion focused against Israel. This was also a case of applying high technology and state controlled Rapid Dominance against a low-technology guerrilla warfare force. Clearly the Hezbolla appeared to win more than they lost in this exchange. The lessons learned from this tragic incident as well as the applicability of Rapid Dominance techniques in this environment need further study. The massing and movement of refugees in large numbers is a reality and a planning factor that must be dealt with up front. The fact that the value of life itself is viewed differently by warring factions must also be considered. If one side willingly uses refugees as a shield and the other is trying to protect their lives, then operations to achieve Rapid Dominance require clear (and perhaps restrictive) rules of engagement in the field. The rapidity of response may not always be the right tactic and an escalation of targeting different centers of gravity rather than responding directly to events in the field promises to be more effective. The theory of Rapid Dominance clearly needs further development, gaming, and simulation. Each decision to apply Rapid Dominance will be unique, complex, risky, and different than the previous one. Knowledge and information on the battlefield as well as that concerning center of gravity targets will be incomplete even with a goal of total situational awareness.

Instruments to Achieve Shock and Awe

Shock and awe are actions that create fears, dangers, and destruction that are incomprehensible to the people at large, specific elements/sectors of the threat society, or the leadership. Nature in the form of tornadoes, hurricanes, earthquakes, floods, uncontrolled fires, famine, and disease can engender Shock and Awe. The ultimate military application of Shock and Awe was the use of two atomic weapons against Japan in WWII. The Shock and Awe that resulted from the use of these weapons not only brought an abrupt end to the war with Japan (through unconditional surrender), but have deterred the further use of these weapons for over 50 years. Not unexpectedly, these events did not stop the proliferation or increase in the destructive power of these weapons by a factor of ten. The holocaust was a state policy of Shock and Awe that stunned the world in its brutality and inhumanity. Yet it has not deterred the world from executing or tolerating atrocities of equal brutality and inhumanity (Cambodia, Syria, Rwanda, etc.). Similar applications of Shock and Awe have differing toleration levels and impacts depending on the environment and political system against which it is applied. As an example, the massive bombing raids of WWII by Germany and the U.S. did not result in a sufficient level of Shock and Awe to end the fighting. The fear of the unknown created by the atomic attacks rather than their actual destruction was the deciding factor in that theater. The B-52 raids in Vietnam provided localized elements of Shock and Awe, but until applied to the capital city of Hanoi, had no impact toward war termination. When applied in concentrated repetitive strikes in November/December of 1972 under *Operation Rolling Thunder III*, the cease fire followed in short order. In fact, throughout history there have been weapons and tactics designed to create varying degrees of Shock and Awe. While there has always been shock, awe, and fear associated with warfare, unless the fear or losses are focused and great enough, a quick cessation of hostilities under favorable terms is not certain. How to apply elements of Shock and Awe against rogue states, terrorist elements, international drug and crime cartels, as well as in the more traditional MRCs and LRCs needs much further study and analysis. Shock and awe, to reach the level required to achieve Rapid Dominance, must also bring fear to those who are in charge. It must be applied quickly, decisively, and preferably with impunity (such as stealth bombing with air superiority). The element of impunity, that is the other side is powerless to stop the damage, is a key element of this strategy. If on the other hand attacks are directed at the general public a backlash could be unleased because of the excessive and brutal losses of innocent civilians.

Much more study and analysis is needed to identify and examine the pros and cons of a policy that initiates a doctrine of Shock and Awe for limited objectives rather than responds in kind to a provocation. What are the limits of the doctrine of Shock and Awe? What circumstances merit the application? Can Shock and Awe be used to achieve limited objectives with little or no risk of life to allied forces or innocent civilians? Can true center of gravity targets be identified for ideological/terrorist groups? Can levels of Shock and Awe be categorized by effectiveness and priority of weapons systems? If so, what are the key enabling technologies? What types of Shock and Awe would be both impressive and generate high returns? A few desirable capabilities from a former CINC's perspective are listed below:

- Blow up an entire mine field simultaneously in its entirety immediately after it had been laid.

- Destroy the mine laden mine-laying vehicles at their loading point.

- Destroy in real time terrorist training camps or publicity generating threats such as the recent display of 70 bomb laden suicide terrorists pledging to wreak havoc worldwide. (This probably requires inside penetration of the targeted organization).

- Destroy simultaneously all/selective WMD launchers, storage/production facilities of a rogue state.
- Selectively target rogue terrorist leaders as was apparently done by the Russians in Chechnya recently when they killed the top rebel leader by detecting and homing in on his satellite phone conversation (helicopter rocket attack).
- Stop, divert, capture the cash flow to terrorist elements.

Thoughts on Applications of Shock and Awe

It is the use of Shock and Awe to achieve Rapid Dominance that is so fascinating and has the greatest potential for leverage if it can be harnessed in a variety of situations. This basis for Rapid Dominance requires a clearer understanding of what our end objectives are than we usually have when we stumble into the use of military force, often it seems by default and at the last possible minute. At this point, I have more questions than answers. How does Rapid

Dominance differ by the goals and missions assigned? What are the key elements to apply Rapid Dominance for each envisioned threat? What are the most likely threats for the next 20 years? Is Rapid Dominance applicable to all these threats? Can we separate Rapid Dominance into categories with and without Shock and Awe?

In addition to answering these and other questions, it seems to me it would be helpful to generate a list of desirable capabilities that would help me select a response option. This list of capabilities would be useful to focus (1) scarce R&D dollars to fill in the holes with technology, (2) intelligence and surveillance collection priorities, (3) innovative thought to further develop the concept (War College papers and Wargaming series), and (4) development of CINC plans and requirements to meet these capabilities. Examples of such capabilities are:

- Deploying highly effective TBMD and Cruise Missile Defense.

- Severing all/selective communications between leadership and field as well as selective elements by call in the field.

- Intercepting and transmitting revised orders to selective threat field units.

- Projecting false radar pictures on selective key threat scopes.

- Inserting fouled fuel in threat storage facilities that generates engine failures.
- Inserting metal/material fatigue to failure attachments on key threat systems.
- Identifying specific location and determining strength and material of protected targets of value.
- Developing dial a setting ordnance capable of destroying all hardened targets.
- Detecting and tracting (destroying at will) all targets of value including mobile targets.
- Detecting and targeting key threat launch systems before launch.
- Detecting plot and simultaneously destroying an employed mine field (land & sea).
- Making threat submarine movements transparent to targeting at will.

Obviously, such a wish list should be prioritized and tailored to the limits of achievable near/mid-term technology and affordability. This may not even be the right type of capabilities one might want. That is, we may need a totally non-standard list. My judgment is that we should develop one or two black "silver bullet" capabilities, if we get too far afield, the system will not be able to digest the recommendations. However, the concept of Rapid

Dominance requires stepping to a new level of getting inside the opposition's decision loop. Rapid Dominance at the ultimate level would enable stopping, diverting, or changing the decision process and decision executing machinery/systems either preemptively or reactively in time to ensure core U.S. security requirements are met.

Rapid Dominance Infrastructure

The current direction and speed of downsizing and acquisition reform is adequate for the type of forces and capabilities necessary to implement a Rapid Dominance strategy. I would like to reserve comments in this area until the project is further developed. We do not need to raise reasons to discard the concept as too hard before it is sufficiently defined. I have the feeling that bringing these conceptual capabilities to realities within a system of systems is neither cheap nor easy. There is still too much waste and inefficiency in our defense acquisition process as well as in the overlap between service requirements and capabilities. Rapid Dominance will not be service-unique and requires a synergistic approach from planning to execution.

Final Thoughts

The implications of the ongoing revolution in telecommunications and information processing as it applies to our national security interests dictate that we need new imaginative concepts of operation to ensure the efficacy of our international leadership in a multipolar world. With technology upgrading capabilities by factors of 10 or more every 18 months, we can no longer afford to have concepts of operations wait for the technology to reach the field. The concept of Rapid Dominance requires innovative thought and different directions than that imbedded in our military hierarchy. We need to introduce the concept at all levels of military professional education and training. The best results of this effort will be generated from the younger minds brought up on the leading edge of the information revolution. The challenge is to engage those minds in the solution and to take the risks required to fund priorities enabling the development of this capability now. Such a cultural change is not easy. One thing is certain—business as usual will not get us there. The window of opportunity will close faster than we think.

Defense Alternatives: Forces Required

General Chuck Horner, USAF (Ret.)

The end of the Cold War will require a review of United States National Security Policy and a concomitant change in our National Defense Strategy. This strategy will respond to the changes in the world's security environment, including the dissolution of the Soviet Union and Warsaw Pact, the evolution in U.S. security alliances such as NATO and NORAD, the increased and unique threat posed by the proliferation of weapons of mass destruction, and the widening of the spectrum of conflict which will challenge the peace and security of our nation and its allies.

The causes of conflict and the modes which threats to our security interests will take have multiplied with the end of the Cold War. The nuclear weapons of the Cold War remain and will remain for some considerable time, even though there is a growing appreciation as to the declining utility of these devices. For sure there will be continuing pressure throughout the world to eliminate the presence of nuclear weapons in conjunction with efforts to halt the production, stockpiling, and deployment of chemical and biological weapons. It is likely that START II will be followed by START III and IV, as nations who claim ownership of nuclear weapons realize ownership has a high cost and marginal payoff. However, progress will be slow due to the immense importance of achieving symmetry during nuclear disarmament and the cumbersome and exacting safeguards associated with the disarmament process. Therefore, for the foreseeable future the threat of nuclear war must be addressed even though it will be less likely than before. The spectrum of national security challenges will expand as the threat of nuclear annihilation subsides.

The decisive victory achieved by the coalition forces over Iraq during *Desert Storm* should give future aggressors of major regional conflict cause to pause. While this does not mean that the threat of conventional warfare has vanished, it does mean that the national leader intending to use major conflict to achieve political aims must carefully craft strategy that will avoid the opportunity for confrontation with a large coalition force lead by the United States. Such a strategy might include surprise attack; short intense military action; the threat or use of nuclear, biological and/or chemical weapons; advanced surveillance measures and precision munitions; and warfare carried out on a fragmented battlefield which includes attacks on the capitals of other nations by means of ballistic missiles or unconventional warfare forces. This will be warfare for which the United States is ill trained and ill equipped.

Other challenges to the world's security will take many forms to which the military forces of the United States can play a constructive role. These are commonly referred to as Operations Other Than War, even though they may include the use of force to achieve desired political goals. They include the increasingly familiar peacemaking, peacekeeping, show of force, and humanitarian relief efforts. Success in these operations may well require retraining, re-equipping, or reorganizing our military forces. Each mission should be evaluated with respect to what is required to accomplish its unique challenges. However, the basic doctrine, training, or equipage of the military forces should be based on what is required to fight the residual Cold War, as well as deal with the growing demands of a major regional conflict.

The political goals upon which our national security strategy should be crafted are fairly straightforward. First, we should seek to preserve and invigorate the role of leadership the United States has maintained since the end of World War

II, or the end of the Cold War (you take your pick). Second, and not apart from the first goal, the United States must be sufficiently strong to prevent or deter use of effective military power against us. It is not inconceivable that our so-called superpower status could be defeated in battle by a crafty and well-prepared adversary. Witness what happened to the powerful victors of WW II in Vietnam. Third, U.S. military forces must be of sufficient size, configuration, and readiness to bring a major conventional conflict to a successful termination. It goes without saying that during this process we need to reduce nuclear weapons to numbers that do not threaten the virtual destruction of the world. Nuclear deterrence forces also must remain in place. Fourth and lastly, our military forces must be capable of responding to all the other tasks and functions for which the national command authority calls upon the military. This first of challenges should be used to define the military forces we field, how we train them, and the methods we use to employ them.

The strategic geographic depth the United States enjoys, bounded by two oceans on the east and west and non-threatening nations to the north and south, means that our nation is somewhat immune from attack, other than by means of infiltration such as a terrorist, or from the skies by means of long-range aircraft, and cruise or ballistic missiles. We will require some actions and defenses which address these threats, but the major portion of our national defense effort must be placed on building and sustaining offensive forces for combat in environments other than our own soil. This dictates that our projection forces must be capable of rapidly responding to an unforeseen crisis anywhere in the world, keeping in mind that quick, decisive surprise favors our potential enemies. Given that we have proven unable to predict the outbreak of conflict in the past, these forces must also be ready at all times to carry out combat operations in most any place. There will not be time to modernize their equipment or train reserve force units. They must be capable of projecting and sustaining their military power over long distances and operating in the environment of the enemy's choosing. Last but not least, when required, they must be capable of decisive combat, not by attrition of the enemy force in head-to-head combat as was our nature in past wars, but by Shock and Awe so that conflict resolution is achieved with a maximum of success at the minimum loss of life in the shortest time. These characteristics for our projection force cannot be achieved easily, as the processes that defined our Cold War doctrines, force structures, equipment, and ways of doing business are loath to change.

The Services' and joint requirements oversight processes that define the equipment provided our military forces place emphasis on force structure and the traditional roles for those forces. This inertia can freeze our land, sea, air, and space capabilities at current or near current levels, but may prove inadequate to carry out new strategies. There are few incentives for a Service or the Joint Staff to reward innovation or divestiture of roles or missions in order to change the character and mix of land, sea, air, and space forces and to prepare them to fight the battles we must envisage for the twenty-first century.

For example, the Services claim lessons learned from *Desert Storm* which reinforce late twentieth century ways of fighting and ignore the troublesome aspects which loom in the future and threaten our traditional view of the battlefield. Many acclaim the role of precision weapons for our forces, but ignore the threat they pose if they are in the hands of the enemy. What would be the lessons learned if several hundred canisters of live Sensor Fused Weapons were released by a red force ballistic missile on the 24th Division during a Fort Irwin engagement? Certainly there would be profound changes in tactics, doctrine, and equipment indicated for the surviving U.S. Army force. What if radar homing Surface to Air Missiles were employed by the red force during a Red Flag exercise in the Nevada desert, not using centralized Soviet tactics/doctrine, but instead using decentralized yet cooperative engagement operations as would be used by our best and brightest if unleashed from their stagnant doctrines? I doubt that the Air Force would be spending millions of dollars trying to build electronic countermeasures to hide the large number of expensive and very non-stealthy aircraft they continue to build, such as the F-15E.

Imagine the shock on our populace if a single cruise missile were actually allowed to score a direct hit on the Carl Vinson aircraft carrier during a Solid Shield joint exercise with the attendant loss of life numbering in the 4,000 to 5,000 range. You would think the maritime force would reexamine the method it provides air power from the sea, vital yet today too vulnerable.

How many times do we hear that the space forces are configured to provide intelligence from overhead only to find in Iraq or Bosnia that the front line forces receive products that are old, inaccurate and altered to keep our Soviet foes from gaining knowledge of our capabilities? Perhaps we if we would dual hat the Director of the Central Intelligence Agency to the position of J-2, or even Commander-in-Chief of a regional unified command, there would be vast improvements in the tasking, evaluation, and delivery of space-derived intelligence to regional combat forces. Then we might see full understanding of the increasing role of space forces and implement change to make them more relevant to our national security strategies of the next century. Innovation, not size, must be sought because we do not have the resources to do both. Moreover, large forces drive our operational level strategy to force-on-force engagements in the attrition warfare model of the last century with its attendant causalities and destruction of equipment. George Patton's dictum still stands that directed his troops not to die for their country, but to get the other SOB to die for his.

Military operations will also place less emphasis on dying and destruction. The ever-present television camera ensures that the horrors of war are broadcast worldwide. War's immorality should some day lead to its banishment. Unfortunately, that day is probably a long way away. Nonetheless, weapons of war and their employment tactics must minimize death and destruction. This is not a call for non-lethal weapons; it is a call for military forces to get right to the heart of the enemy and conclude operations as rapidly and efficiently as they possibly can given their equipment, training, and doctrine. This means there must be wide flexibility in how they may function. Military operations will be across a wide spectrum of warfare and will demand flexibility. Modern war will require our military leadership to navigate through a changing spectrum of political constraints and ever changing political goals as each scenario unfolds. We must make our forces capable of dampening the capacity of the enemy to use force by controlling the conflict rapidly even when surprised. We failed to do that tactically in *Desert Storm* in the case of the SCUD missile attacks, but were fortunate that the Iraqis were equally inept at taking political advantage of this card they held and skillfully employed on the battlefield. We must also look for efficiency before we even join in battle.

Defense spending has declined as a percent of federal outlays since the end of the Cold War. Given the leadership role the United States plays in the world, one could think a reasonable sum to devote to defense might be three percent of our gross national product, certainly an amount much smaller than what an average family expends for its security by means of life, health, causality, car, medical insurance, and retirement benefits. Given the prospect of long-term, constant funding, the Department of Defense could then give more thought to how to build the most modern, efficient military force within the dollars available. We would no longer define our forces against some mythical threat or scenario which generates impetus to protect force size rather than quality. The Army, Navy, Air Force, Marine Corps, and space forces would be required to build a team based on a salary cap. You might be willing to pay big bucks for a B-2 superstar quarterback, but you will also need lower cost and capable riflemen or destroyers to block and tackle. Most of all, you would reward the Service or Agency who would innovate to provide efficiency.

Manpower has become the driving cost in the all-volunteer military force. Investment cost of a ship, tank, aircraft or satellite might be high, but it is the operations and maintenance costs that will drive how much resources we are required to expend to gain and maintain a given military capability. Again turning to *Desert Storm*, the huge advantages of overflight precision munitions dropped from stealth aircraft has not been understood or accepted by the operations analysts who argue what we should build or buy next. If it had been, would the Navy have allowed the A-12 program to fail, would the Air Force be pouring hundreds of millions if not eventually billions of dollars into equipping forty year old B-52s with conventional missiles, or would the Army be maintaining heavy divisions at a personal cost of $60 billion for 35 years of ownership? Why not build a Division force equivalent using technology and doctrine to provide a "heavy division equivalent" force using far fewer troops featuring speed, shock, precision fire while avoiding the manpower costs of dollars that in peacetime include added costs for recruitment, training, and sustaining and in war have an even greater added cost computed in blood? Why don't we do this? The answer is because it would require rare innovation, trust, and support from the equally intransigent federal funding authorities. Most importantly, the Services are not rewarded for innovation which recognizes the contributions of another Service or Ally.

Jointness has become an altar at which all military personnel must worship even if they don't understand or believe. Defenders of the status quo argue that there is merit in duplication or redundancy and these arguments have some validity. The question becomes how much overlap or redundancy between land, sea, air, and space forces can the nation afford, and what is the opportunity cost to the core competency of the land, sea, air, or space force that builds and/or maintains the duplicative force structure. A second yet vastly different question arises when considering the unique capabilities a Service provides to support itself and the other services. For example, how much the Air Force should spend on airlift forces is not cast in terms of what the envisaged requirement is for airlift, ton miles per day, to support the mythical scenarios. The alternative sea, land, and space lift requirements can be postulated; however, if the Navy, Army, or Air Force do not satisfy those sea, land, and space lift requirement, then there is a shortfall which will in turn generate a need for more airlift!

During *Desert Storm*, nearly 90 percent of the deployed equipment arrived by sea, but not in time if the Iraqis had continued their first attack in August. A majority of overland movement was provided by Saudi Arabian civilian trucks and drivers, and the Army had neither the resources nor the responsiveness to activate reserve forces needed to meet the truck and rail support requirements of our military forces. As a result, costly airlift was used to move forces that should have traveled by land and sea. If added space capabilities had been needed, there was almost no capability for the timely launch of a satellite. Would it not be wise to index spending on land, sea, air, and space launch on one and other, postulate lift requirements on what the new force needs as it innovates and slims down. The need to respond on a moment's notice adds to the value of airlift and prepositioned ships. The outcome though would be not to allow any of the Services to divert general support money into core competencies and thereby shift the jointness burden to another Service.

Innovate. Use the carrier to haul the army to war, and then fly the fighters aboard after the helicopters or tanks are unloaded. Accept the benefits of Federal Express that can be federalized during times of national emergency as a costly, but ready augmentation to military supply lines that has no cost during the much longer periods of peacetime. Our nation has other industrial capacities that also have duplicate military capabilities. They may be 80 percent solutions, but the cost of ownership could prohibit creation and maintenance of a military owned and operated 100 percent solution. Iridium telephones may not be jam-resistant or secure, but 80 percent of the time they will satisfy the need for 2 percent of the cost. Of course, this avoids the problem we have created for ourselves with our medieval acquisition system.

Finally, we must acquire hardware of a type and at a pace that will assure the future force capability will be enduring. We cannot keep up with technology using our current ways of acquiring military hardware and training our people in how to use and maintain it. In many areas we would be better off to throw it away when it breaks given the low cost, durability, and reliability of modern solid state electronics. Why train technicians? Give the troops a gold card and a telephone number and they know how to spend money more efficiently than do our government agencies. Make sure the equipment we do buy not only integrates with that of other services and functions, but that it can integrate with both older and newer equipment designated to do the same function. The fighter aircraft secure radio must be capable of communicating with the ground and sea based forces command and control, as importantly it must be able to communicate with the next generation fighter aircraft radio.

The added dimension is the realization that we are unlikely to fight alone in the future. We gain valuable legitimacy from forming coalitions, plus it makes up for the growing feeble force structure we maintain in declining budget years. An enduring force must also recognize the necessity to operate cooperatively with the forces of other nations. This means we must more freely release our technologies to foreign nations so that our military forces can fight side by side, so that our deployment forces can draw from stocks of others while our logistics system seeks to catch up with the rapidly deployed combat force.

In the final analysis, all of this shaping and sharpening of our military forces will be for naught if there is not an equal change in the policy side of the equation. What good are highly trained, efficient, capable land, sea, air, and space forces if the implementing authorities are incapable of defining principles, goals, and integrating strategies for their employment? While this is not the province of the military to solve, the military must understand how disjointed policy, weak political leadership, or dysfunctional international cooperation will preclude success on the battlefield.

Again, one of the missed lessons of *Desert Storm* was the difficult and successful integration of international leadership achieved by the President, Secretary of Defense, Chairman of the Joint Chiefs of Staff, Congressional leaders, and allied National Command Authorities as well as many others. It was this leadership, coupled with the ineptness of the enemy, that covered over the failures of our Cold War-equipped and trained forces that fought *Desert Storm*. This does not take anything away from the military victory, but it does make it difficult to glean the right lessons for the future. Perhaps that is why we are so loathe to change our forces at a time when change is demanded by a new strategic environment and new threats to our national security. Defining alternative forces in light of the changed national security environment, goals and strategy raises two questions: what kind or mix of military force and how much best balances the requirements and funds available.

Deep Strike: A Key to Shock and Awe

In the world of surprise attack and withdrawal from foreign bases, all initial responses to combat operations will be some form of deep strike. Given strategic warning (don't bet on it) after deployment of our military forces, Deep Strike is a term that relates to the political boundaries or proximity to military forces. The geography of the area of conflict will further define deep strike. But a rule of thumb might be attacks on a target beyond range of surface-based fires except for ballistic or cruise missiles. More important than range is the characteristics of the Deep Strike targets. Deep Strike targets could be classified as ones the enemy does not wish to place at high levels of risk. They can be characterized by the functions they perform, such as:

- Leadership
- Command and Control (a function of leadership)
- Control of Military Forces, especially air and space
- Logistics and Sustainment
- National Economic Base
- Internal Security/Political
- National Will, Theirs and Ours

Intelligence used to nominate the targets for these strikes must examine the functions and then define the physical objects or people who comprise the system which is responsible for the successful operation of the function. You define the system and then attack the critical elements in order to achieve economy of force. Often these target sets are difficult to define, as these functions often represent the enemy's most valuable and therefore protected elements. The intelligence collection associated with each function will vary from target set to target set. Large, fixed infrastructure, such as associated with an electrical grid, lends itself to traditional reconnaissance and evaluation of technical analysis. Leadership targets are better defined by using human intelligence and subjective analysis. In all cases success starts with innovative intelligence products, which has not been a hallmark of United States operations. Such intelligence products must be examined through the eyes of the enemy, their values and concerns. Too often we apply judgments based on our viewpoint.

One target system may serve the attainment of a number of different goals. For example, attacks on the electrical power system of the enemy may debilitate his capacity to command and control his military forces, operate vital elements of the economy and thus degrade the political support required to sustain the conflict. This same target system may be attacked a variety of ways. Most common methods would be using stealth aircraft and cruise missiles to bomb power plants and switching centers. Areas with isolated populations lend themselves to using special operations forces infiltrated to destroy an isolated power grid node for transmission of energy from one highly populated area to another. Now it is obvious that computer signals used to command the power grid are targets as intrusion into the enemy's control system provides the means to simply turn off electricity to selected areas. Attacks by all these means achieves even greater results than the sum of its parts because enemy responses to restore electrical power will be confused as elements such as computer intrusion are confused with bombing destruction.

The characteristics of value in attacking these important targets systems are simultaneity, impunity, and timing. The greatest effect will be achieved when the strikes are coordinated in such a manner as to inflict maximum Shock and Awe on the enemy element. This means operations must be coordinated and orchestrated carefully and flexibly as enemy reaction to the attack is evaluated. Moreover, presence is projected when a combination of functions or target sets supporting a variety of functions are struck at the same time with impunity. In order to achieve maximum results, the attacks will need to be evaluated quickly in order to define previously unknown elements of the system or how the enemy perceives the impact on his system.

Finally, the attacker must be alert as to the interaction of the functions as the effects of these Deep Strikes begin to take hold. In order to achieve desired levels of Shock and Awe, the attacker must know the current and projected effects of his strikes against elements of the enemy's residual system. If the trick is to define the system of targets needed to conduct successful Deep Strike, it is even more important to know how to alter the initial plan as the battle unfolds and timing becomes everything.

The characteristics of forces needed to carry out Deep Strike are long range, flexibility, precision, survivability, and speed. Cost of the operation is a factor; however, system cost must include peacetime operations and maintenance costs as well of the costs during actual combat. There is also a human element in the cost of combat operations which escalates rapidly as military force is misused. The total cost of these operations must also address the cost of intelligence used to support Deep Strikes. Intelligence operations may be the most costly due to the importance of these targets to the enemy. Alternatively, the human intelligence associated with these attacks may be the most inexpensive since their national importance makes them vulnerable to knowledgeable dissidents.

Stand-off

Deep Strike is defined by distance, albeit relative distance. Some of the target sets may lend themselves to circumstances beyond the nation's control; for example, Seoul borders on North Korea. Our protective oceans mean that likely conflict is offshore. The likelihood our next adversary may have access to surveillance, precision munitions, and long-range delivery systems dictates that much of our operations will be at long range, lest our forces come under attack at their ports, camps, and bases. There will be a need for systems capable of projecting military force from distances of 10,000KM. A sizable portion of the force must be able to deliver ordnance of enemy targets from ranges in excess of 5,000KM. Launching attacks from inside 1,000KM of the enemy forces will demand that friendly forces be protected from attack by means of active and passive defenses and dispersal. This latter constraint will preclude achieving levels of Shock and Awe through simultaneous attack.

Survivability

Great cost benefits are attained if the vehicle used to deliver the attack is reusable. Keep in mind that the force built for the most demanding conflict must also be flexible for other operations. Therefore, while ballistic missiles provide great range, speed, and survivability in reaching their target, their cost become prohibitive in large-scale operations which endure beyond a few hours, or in smaller-scale operations where the goals are modest and the demands on other military forces are low. Simultaneous combat operations require a number of expensive, expendable platforms in the opening hours of the conflict if our response is to be timely and induce shock. Awe is not achieved if the enemy is permitted to gain experience in being attacked; at best you may make them numb. Alternatively, reusable long-range survivable systems provide needed flexibility to alter the Deep Strike plan as it unfolds. The food chain of weapons systems ranges from the most valuable systems such as ballistic missiles, cruise missiles, and stealth bombers, to less valuable, but useful, stealth fighter and long-range surface-to-surface high trajectory fires.

Firepower

Discriminate fires are important due to the likelihood of people and structures being in close proximity to the desired target. It is not improbable that the national command center is located next door to a children's hospital.

Discriminate fires require precision in target coordinate identification and location. Precision does not mean "small warhead," although there is a beneficial impact as the right amount of explosive is placed on the target due the penalties imposed on the delivery vehicle required to carry the warhead long distances. All operations involving the use of firepower must also understand and evaluate the beneficial aspects of using non-destructive elements in conjunction with the attack to include all aspects of the so-called information warfare.

Enduring Realities and Rapid Dominance

General Fred M. Franks, USA (Ret.)

Rapid Dominance, as we see it, is a markedly different concept for the use of force to gain national security objectives. At its core, Rapid Dominance blends unique capabilities of land, sea, air, space, and special operating forces. It is important to note the vital role of jointness in using forces from all elements and resisting the lure of gimmicks and cost-free options that may appear within the reach of high technology but are not.

Examining current joint force capabilities reveals some enduring truths that should be used to evaluate future concepts. Joint force commanders today benefit from the wide array of capabilities available to the joint warfighting team. The ability to combine and use forces from all dimensions in a variety of powerful combinations to fit mission circumstances presents a versatility of capabilities that makes defense by adversaries difficult. Balance and versatility are key. Balance in capabilities and the inherent versatility to combine them in unpredictable, yet highly effective ways has served U.S. national security interests well since the end of the Cold War. One has only to look at the variety of methods employed in Panama (1989), *Desert Storm* (1991), Somalia (1992), Rwanda (1993), Haiti (1994), and Bosnia (1995) in both war and operations other than war. Joint force commanders employed, and in some cases invented, new combinations of balanced capabilities and were willing to go beyond the confines of service doctrines to fit mission circumstances. For example, a U.S. Army brigade of the 10th Mountain Division with helicopters replaced much of the carrier air wing and flew off the carrier *Eisenhower* during the Haiti operation. This force packaging capability is an advantage unique to the U.S.

As we look beyond the present to future and bolder defense concepts such as Rapid Dominance, the key will be to maintain that balance in land, sea, air, space, and special operating forces combinations available to the joint force commander. U.S. military forces are now multidimensional in capabilities, able to use force in ways unpredictable to an adversary. U.S. forces also have enormous versatility, able to be used in war and what have become termed operations other than war. Balance permits that.

Moreover, joint force commanders, recognizing this capability, have found ways to introduce land forces even more rapidly given today's methods. Recently, a brigade of the 1st Cavalry Division rapidly deployed by air from Ft. Hood, Texas, to Kuwait and was able to fall in on equipment forward positioned and be available for combat soon after arrival. A recent article in *Navy Times* pointed out, "In fact, as each wave of soldiers arrived in Kuwait, they were heading north—combat ready—within six hours." This was a dramatic example of the rapid ability to combine land forces with air and sea forces using both distant forces with those already in the theater. That combination in that set of strategic circumstances provided a rapid deterrent in an area of vital national security interests to the U.S.

Another enduring truth is the need for staying power and ensuring that capacity is perceived by a potential adversary. Staying power means the ability to press the initial advantage gained until the strategic objective is achieved. On-the-ground presence, in addition to forces in theater, as demonstrated in Kuwait in 1993 and again in 1996, provided commitment and staying power to convince Iraq that it would be disastrous to consider any form of military action. The inherent staying power of land forces, wherever future tactical concepts may lead, makes them a powerful contributing partner in our Rapid Dominance concept.

Finally, there is the issue of physical control. Control combines with staying power to defeat the enemy's will. One of the many lessons of *Desert Storm* is that it was not until after land forces attacked Iraq and Kuwait that Iraqi forces were expelled from Kuwait. Despite the awesome shock and destructive effects of attacks from the air and sea, it was only after coalition ground attacks to extend control to both Kuwait and southeastern Iraq by defeat and destruction of defending Iraqi forces that strategic objectives were secured. Control on land was extended past the cease fire until such time in April as the UN passed a permanent cease fire and sanctions resolution. Land forces remaining in southeastern Iraq provided the staying power and control.

The size, shape, and composition of forces that will fight in all elements will assuredly change in the future. Early work done in advanced warfighting experiments out of TRADOC's Battle Labs beginning in 1992 and growing into the current Force XXI and other promising capabilities as well as by the USMC at MCCDC at Quantico are the precursors of how change may be discovered and implemented. The challenge is to ensure that all components of our fighting power are properly balanced and combined into the most effective and lethal mixes of land, sea, air, space, and special operating forces. This is the heart of the Rapid Dominance force of the future.

Extension of real and perceived control over the will and ability of any adversary to oppose or threaten us will insure and guarantee success of initial operations, thereby maximizing Shock and Awe. Indeed, getting forces on land rapidly and operationally will be a major factor in achieving the enduring effects of Shock and Awe. Certainly, as forces on land evolve and change, they must meet the requirements of rapidity and sustainment and are vital components of any mix of forces that seek by Shock and Awe to stun and then rapidly dominate an adversary to achieve U.S. national security objectives.

We strongly feel that we as a nation cannot stand still in exploring defense alternatives. We must seize this time to be bold in our thinking. More thought and hypotheses with operational methods that break through or expand current service doctrines are needed from a joint perspective even as services look to the future from their own service perspective. Then there must be rigorous experiments using both high fidelity simulations and actual joint field trials to determine the worth of these hypotheses to blend the wide array of technology available to the total joint force and according to bold new concepts. The results will determine the worth of Rapid Dominance concepts by judging whether they will permit even more balanced, versatile, and lethal combinations to fit known and anticipated future strategic circumstances.

Study Group Members

L.A. "Bud" Edney is a retired Navy admiral and naval aviator. A veteran of over 350 combat missions in Vietnam, Admiral Edney's senior billets included Vice Chief of Naval Operations and Commander-in-Chief, Atlantic Command/Supreme Allied Commander, Atlantic. Admiral Edney has an advanced degree from Harvard and was a 1970 White House Fellow.

Fred M. Franks is a retired Army general and a highly experienced combat armor officer. During the Gulf War, he commanded VII Corps and last served as Commanding General of the Training and Doctrine Command. He has two master's degrees from Columbia and is a graduate of the National War College. He is the author of *Into the Storm, a Study in Command,* written with Tom Clancy to be published by G.P. Putnam's Sons in 1997.

Charles A. Horner is a retired Air Force general and a highly experienced combat fighter and attack pilot. During the Gulf War, General Horner commanded all allied air forces. His last assignment was Commander-in-Chief, Space Command. A graduate of the National War College, he now serves as consultant to government and industry.

Jonathan T. Howe is a retired Navy admiral and both a submarine and surface warfare qualified officer. He has served as Deputy Assistant to the President for National Security Affairs, Deputy Chairman of NATO's Military Committee, Commander-in-Chief Allied Forces Southern Europe/CINC U.S. Naval Forces Europe, and was Special Representative of the Secretary General of the UN to Somalia. He has a Ph.D. from the Fletcher School of Law and Diplomacy and currently heads a charitable foundation.

Harlan K. Ullman divides his time between the worlds of business and public policy. A former naval person, he is with the Center for Strategic and International Studies and the Center for Naval Analyses. His last book, *IN IRONS: U.S. Military Might in the New Century,* was published by the National Defense University Press in 1995.

James P. Wade, Jr., a scientist by training, is a West Point graduate and infantry officer. He has held many senior positions in DOD, including head of Policy Planning, Assistant to SECDEF for

Atomic Energy, Assistant Secretary for Acquisition, and Acting Head of Defense Research and Engineering. He is Chairman and CEO of DGI which conducted this study.

Keith Brendley is a Vice President with Defense Group Inc. He was formerly with Sarcos Research Corporation, RAND, System Planning Corporation and NASA, Ames Research Center. He holds mechanical engineering degrees from the University of Illinois (B.S.) and the University of Maryland (M.S.).